PRESENTATION

This book documents something that, in my judgment, could be the most important research and discovery in scientific affairs that has been made in Mexico in the last decades. The human photosynthesis' finding is the result of an intense commitment in the search of new medical therapies which brought Dr. Arturo Solís Herrera to question thesis once considered as fundamental truths about life and to bring forth new and amazing theories about it, theories that every day gain more and more approval among the scientific community, both national and international.

To live and to work in Aguascalientes Mexico, far away from the large research centers of the country and of the world, hasn't inhibited the innate brilliant spirit that has prompted Solís Herrera to dig in the boundaries of human knowledge – the same spirit which motivated, albeit in different circumstances, Galileo Galilee, Isaac Newton and Charles Darwin to question the most deep-seated concepts of such knowledge and change history's path – and today again forces us, by the strength of powerful evidences, to modify our ideas about the origins and processes of life.

The finding that not only vegetables perform the photosynthesis, but also human beings and every living being whose genetic code expresses melanin, compels us to rewrite countless pages of science, while opening, at the same time, another infinity of new pathways for research in physics, chemistry, energy, medicine and all basic sciences.

His honest and constant work inside his ophthalmology lab, with tiny photo electrochemical cells, totally redefines our conception on how light acts upon and inside the human body. It shows, without any doubt, that melanin – a little known substance, discarded as useless and irrelevant by many scientists – allows, trough photolysis, to extract energy by binding and unbinding hydrogen atoms from oxygen inside the water molecule. In other words, melanin is fundamental to life, because water molecules, when in presence of melanin and light, turn themselves into hydrogen and oxygen atoms, plus energy, and then, the atoms come back together, giving water and electricity, without changes in the melanin molecule, this allows this reaction to continue through time. This amazing discovery, patented by Solis Herrera, gives a complete new vein in terms of renewable and limitless energy.

Scientific researches in other countries tend to confirm the finding of human photosynthesis, without showing the same insight on the phenomenon that Solís Herrera has showed. A researcher team from the Albert Einstein College of Medicine, lead by Dra. Ekaterina Dadachova has studied the unusual growth of fungus in the presence of melanin and radiations, back in Chernobyl. This fact created speculation in scientific media in the U.S. about the possibility that photosynthesis trough melanin could be the basis for massive cultivation of fungus which could be used as the biofuels of the future.

"*This suggests that nature herself has produced another totally unexpected scenario of alternate energy (...) Does this means that someday big fungus farms in the hills of Los Andes or the Himalayas will give us fuel for our automobiles?"* asks a paper from Technology Review, from the Massachusetts Institute of Technology (MIT).

The photoelectric cells shows a more appropriate option to capture the energy produced by the melanin photosynthesis, to such an extent that its application is already being planned for use in public lightning in Mexico.

It's foreseeable that the future demand for melanin's photo electrochemical cells, both for public lightning and as energy supply, could reach overwhelming proportions. Besides, it's an energy form compatible with the search of ways to fight global warming. Today's unsustainable economic growth is still based in growing consumption of fossil fuel, at the cost of the planet's life systems, thus, radically different energetic solutions are required. Is on this context in which we must understand the true potential of photosynthesis through melanin's power.

David Shields
Director and Founder
Energía a Debate Magazine

The body don´t read books and really all we needed to do were "read" our body.

Abstract in regards the main events that uphold the concept of the existence of photosynthesis in humans.

THE HUMAN PHOTOSYNTHESIS

Arturo Solís Herrera MD, Ph.D.

MELANIN

The human chlorophyll,
Its role in the origin of life
And the possibility that
It may be the much-searched
Universe's Dark matter.

ISBN 978-0-7414-8420-8

Printed in the United States of America

Published April 2013

INFINITY PUBLISHING
1094 New DeHaven Street, Suite 100
West Conshohocken, PA 19428-2713
Toll-free (877) BUY BOOK
Local Phone (610) 941-9999
Fax (610) 941-9959
Info@buybooksontheweb.com
www.buybooksontheweb.com

At the beginning, there was darkness.

Genesis 1:2

Human photosynthesis is a model capable of accounting for all functions of animal cells and in simple terms can be schematized as follows:

The Solis-Herrera Cycle

$$2H_2O \leftrightarrow 2H_2 + O_2 + 4e^-$$
$$\downarrow$$
Energized state
$$\downarrow$$
Work performances

After the unraveling of the Human Photosynthesis, the metabolic models based in glucose oxidation as source of energy must be rethought.

CONTENTS

INTRODUCTION

The whole purpose of this book is essentially to describe and thus gives diffusion to the knowledge about the important effects that the intrinsic property of melanin to split and reform the water molecule has upon the cellular biology, the fact, unnoticed until now, that at room temperature and using natural or artificial light as its only energy source, melanin, both in vivo as in vitro, have the ability to break the water molecule, obtaining diatomic hydrogen and oxygen, and also can achieve the opposite event, to gather hydrogen and oxygen molecules, obtaining water and as well as high energy electrons and therefore electric current, thus, to achieve the former, nature uses melanin as unsuspected electrolyzing material. These reactions occurring both inside and outside the cell, even in vitro, and can happen with more or less regularity, under a thousand possible stimuli, whether physical, chemical, both external and internal.

In other words, melanin's effects are so important to nature that cells begins to raise its activity to synthesize this molecule when any reaction, either external or internal begins to occur in a more or less repeated way. Inside eukaryotic cell, more accurately in the melanophores, the main material, the indispensable solute for the electrochemical principle works upon are soluble melanin, given its notable ability to capture photons in wavelengths that goes from 200 to 900 nm of the electromagnetic spectrum, but not only from this part of the spectrum for the pigment is sensible to almost all the electromagnetic spectrum, something that is probably made by the peripheral sections of the molecule, both in vivo as in vitro, which is followed by a high energy electrons generation arising from low energy electrons. These high energy electrons go to the compound's free radical centers where they are probably captured by some element, for instance any metal, such as iron, copper or any other, where they are transferred to a primary electron taker, of unknown nature till today, because the union is very complex and it comprehends ionic interactions, depending from the pH. This electrons transference releases energy, which is used to set a proton gradient.

Water and melanin's molecules combination forms what can be called a photo system, which absorbs luminous energy, using it to, at least, two intertwined actions: to remove electrons from water and to generate a protons gradient. Melanin components are in very close contact one with another, which makes a fast energy transference. In 3 picoseconds after being lighted up, melanin's reaction centers responds sending a photo excited electron towards the primary electron taker.

This electrons transference generates a positively charged giver and a negatively charged taker. The formation of two oppositely charged species importance becomes clear when we consider the oxidize-reduction abilities of these two species, since one of them is electron deficient and can take in electrons, which makes it an oxidant agent. In the other side, the other compound has an extra electron that can be easily lost, making it a reducing agent. This event – the forming of an oxidant agent and a reducing agent from light- takes less than a second's billion part and is the first step in water photolysis.

Due to being oppositely charged, this compounds show an obvious attraction to each other. The charge separation is (probably) stabilized by their own movement, at opposite sides of the molecule, being the negative compound the first to give up its electron to a quinone (Q1), and, possible, then this electron is transferred to a second kind of quinone (Q2), which produces a semi reduced shape quinone molecule, that can be strongly tied to melanin's reaction center. With every each single transference, the electron gets nearer to melanin's reaction center. The positively charged part of melanin is reduced, which prepares the reaction center for another photon's absorption. A second photon's absorption sends a second electron alongside this path (negatively charged melanin towards first and second quinone molecule –(Q1 and Q2), this second molecule absorbs two electron, and combines with two protons.

Protons used on this reaction could come from melanin molecule itself or from the surrounding water, causing a descending in the photo system's hydrogen ionic and molecular concentration, which helps to the forming of a proton gradient. Theoretically, the reduced quinone molecule dissociates from the reaction center, being replaced by a new quinone molecule. These reactions occurs at room temperature, but could be modified, for instance, by the heat, that can help reactions in one way or another, depending the control we have on the other known and unknown variables, such as pH, magnetic fields, movement, solute and solvent concentrations, intensity of electromagnetic radiations, the latter's nature, melanin's purity, if it's doped or not, gas partial pressures, container's shape, etc.. And the final purpose we want the process to have, according the kind of cell, tissue, or water available. For instance in presence of metal, such as bore, hydrogen works with -1.

Separating the water molecule in hydrogen and oxygen both diatomic is a highly endergonic reaction due to a very stable association between hydrogen and oxygen molecules. The water molecule's separation (in diatomic hydrogen and oxygen) inside the laboratory requires the use of a strong electric current or rise the temperature almost to 2,000 °C. This (water electrifying or photolysis) melanin accomplishes easily both in vivo as in vitro at room temperature, using only energy from visible and invisible light, and not only from a 200 to 900 nm wavelength, whether from a natural or artificial source, coherent or not, focused or disperse, mono or polychromatic.

The estimated red-ox potential of quinone's oxidized form is approximately +1.1 V, which is strong enough to attract the strongly tied low energy electrons from the water molecule (red-ox potential +0.82) which separates the molecule in hydrogen and oxygen atom. Water molecule separation by photo pigments is called photolysis. The forming of an oxygen molecule during photolysis, requires, or so is thought, the simultaneous lose of four electrons from two water molecules, according to this reaction:

$$2H_2O \leftrightarrow 2H_2^+ + O_2 + 4e^-$$

A reaction center can only generate one positive charge, or its oxidant equal, at the time. This problem is hypothetically resolved with the presence of 4 nitrogen atoms inside melanin's reaction center, each one of these transfers only one electron. This nitrogen concentration stores perhaps four positive charges transferring four electrons (one at the time) to the nearest quinone molecules. The electrons transference from the reaction center's nitrogen to the quinone⁺, is accomplished by the passing through a positively charged quinone leftover. The electron, after is transferred to the quinone ⁺, is able to regenerating quinone in the process, then the pigment is oxidize alone (again into quinone+) after another photon has been absorbed to the photo system. So, the storing of four positive charges (oxidant equivalents) by nitrogen atoms from the reaction center is modified by the successive absorption of four photons by melanin's photo system. Once the four charges have been stored, the oxygen releasing quinone complex can catalyze the removal of 4e⁻ from 2EbO, forming an O_2 molecule, and regenerating the completely reduced nitrogen store from the reaction center.

Protons produced by photolysis are released where they are integrated into the protons gradient. The photo system must be lighten up several times before there is an O_2 releasing, and hence, of H_2 also, which could be measurable, which indicates the individual photoreaction's effects must be stored before the O_2 gets released.

Quinones are mobile electrons carriers. We must not forget all electrons transfers are exergonic and occurs as electrons get translated to carriers with a growing affinity to electrons (more positive redo potential). Mobile electrons carriers' necessity is clear. Photolysis generated electrons can go to several inorganic takers, which get reduced by it. These electrons paths can lead to (according the used mix's composition) an eventual reduction of the nitrate molecules (NO_3) into ammonia molecules (NH_3), or sulfates (SO_4^{-2}) into sulphydriles, mercaptants or thiols (R-SH), both reactions turn inorganic wastes into compounds useful to life. So, sunlight energy can be used not only to reduce carbon most oxidized form (CO_2), but also to reduce the most oxidized forms of nitrogen and sulfur, both in vivo as in vitro, with obvious advantages. The former lines was cited more deeply and focused to the in vitro form in patent GT/a/2005/0006.

The production of an O_2 molecule requires removing four electrons from two water molecules. Removing these four electrons requires absorbing four photons, one for each electron. On the other hand, there are also evidences of melanin capturing kinetic energy, turning it into useful energy for the cell, perhaps as hydrogen and oxygen, or the contrary, water and electricity. cytosol and cell membrane's composition are important parameters to obtain this particular reaction's outputs, given the fact that electrolytes' presence, both in vivo as in vitro, its nature, using magnetic fields, using kinetic energy besides electromagnetic radiations, the addition of several compounds – doping – (both organic and inorganic, ions, metals, drugs) into the photo system, which initially only consists of water and melanin, plus electrolytes, as well as temperature handling, gases partial pressures' control, the electric current handling, magnetic fields application, pH levels, makes the final destination to recover electrons, protons or oxygen, as well as resulting compounds, according to the medium in which melanin is dissolved composition. So, the heart of an efficient photo electrochemical design is melanin, both in vivo as in vitro. Electron's transference releases energy, which is used to set up a protons gradient.

Proton's movement along the electrons carrying can be compensated by another ions movement, which are present in the intra or extracellular medium. Melanin's electrolyzing properties (among many others) in vivo, can explain the light generated pike, observed in an electro-retinogram, because when we light up melanin, intracellular pH descends so the H_2 and ionic increases, which activates chlorine channels sensible to pH inside the base-lateral cell membrane. Light's pike is a rising in potential, following the fast oscillating through phase FOT, it forms the slower and longer component of direct current electro-retinogram (Kris 1958, Kolder 1959, Kikadawa 1968, Steinberg 1982).It also intervenes in aqueous humor's dynamics, because they can explain congruently why intraocular pressure descends during daytime, despite the 45% rising of aqueous humor during day, both processes during light's presence, at night aqueous humor production also descends 45% and intraocular pressure rises 40%, so we can observer light energizes notably vital processes of aqueous humor dynamics, which can be supported observing that drugs which stimulate melanocytes indirectly (for instance beta blockers) aren't effective by night. Melanin, melanin's precursors, melanin derivatives, melanin variations and analogues, remove electrons from water and generate a protons gradient.

Light's dependent reactions can also supply energy to reduce CO_2 into CH_2O, nitrates to ammonia and sulfates into sulf-hydriles, both in vivo as in vitro. Melanin absorbs all spectrums of electromagnetic radiations, even hard and soft ultraviolet, all visible and invisible spectrum and both near and far infrareds (Spicer & Goldberg 1996). It's not too farfetched that it can also absorb energies like kinetic or other wavelengths far away in the electromagnetic spectrum like gravitons. It is very interesting to consider using melanin's photo-electrochemical properties on industrial processes based on biological systems, like generating hydrogen and oxygen or an alternative generation of electric energy.

CHAPTER 1

Brief Story of Photosynthesis in Plants

Photosynthesis is a term that describes how the plants are able to combine the CO_2 with water (H_2O) to made Glucose ($C_6H_{12}O_6$) using the energy of the light in order to impel this not yet well understood process. In the literature we could found other definitions as follows:

Photosynthesis: "High relevance process through which green plants capture de sun's visible energy to induce the generation of energy (ATP) and reductive power (NADH) both elements are necessary for the synthesis of carbohydrates (as glucose)".

A cycle of carbon-energy is described, consisting in a part of it being taken by animals, which degrade glucose in presence of oxygen, generating carbon dioxide (CO_2) and water (H_2O). However, this part of the cycle is not self-sustainable and photosynthesis is the part of the cycle that allows the reuse of the CO_2 and water to rebuild the carbon hydrates. The reverse of oxidation process from animal metabolism (and hence of human beings) over the carbon hydrates is completed by plants, algae and some microorganisms, using the sunlight's energy to provide the huge amount of energy required. As we already established, this process is called photosynthesis, and not only does it provides carbohydrates for the energy production, but is also the main path through which carbon goes back into the atmosphere, this means that it's the main way to hold carbon, as well as the main source of oxygen in Earth's atmosphere.

In the actual process of photosynthesis, there are many in between steps. Besides, and as hexoses are not the main source of carbon obtained, thus, the photosynthetic reaction is written generally like this:

$$6CO_2 + 6H_2O + \text{Light Energy} \rightarrow C_6H_{12}O_6 + 6O_2$$

But this is merely a schematic reaction, for the general reaction.

All aerobic organisms produce CO_2 when they oxidize carbohydrates, fatty acids, and proteins in the mitochondria of cells. The large numbers of reactions involved are exceedingly complex and not described easily.

However, in photosynthesis the very first and the one we have interest right now is the following:

$$2H_2O \rightarrow 2H_2 + O_2$$

This is the very first step in plants photosynthesis and is considered the most important reaction in the world because is the starting of the food chain.

At the same time, this initial reaction, which means photosynthesis (build up something with the force or energy of light) starts with the dissociating, unfolding or breaking of the water molecule, something that, obviously, needs energy to be done, because it cannot happens spontaneously. For it to happen, energy has to be applied and this energy the plant takes it from the sunlight through chlorophyll. As such, it's the only step in which the chlorophyll molecule steps in and only to capture the photonic energy, processing it in a way not entirely understood so far, to break the water molecule. In other words, it separates hydrogen from oxygen (chlorophyll oxidize irreversible the water molecule) and immediately puts the former (diatomic hydrogen) at the disposition of the plant cell, which, with the aid of other molecules continues the synthesis of carbon hydrates. However, chlorophyll has nothing more to do with it. Its role ends once it unfolds the water molecule.

From the here above description, we want to outline these words about photosynthesis: "High relevance process".

As part of the carbon cycle known as photosynthesis, plants, algae, and cianobacteria absorb carbon dioxide, light, and water to produce carbohydrate energy for themselves and oxygen as a waste product.

In accordance with our unraveling of human photosynthesis process, carbohydrates are not a source of energy, instead are source of biomass.

Respiration is the opposite of photosynthesis, and is described by the equation:

$$C_6H_{12}O_6 + 6O_2 \rightarrow 6CO_2 + 6H_2O + 36 \text{ ATP mol}$$

But at light of our research this is not plausible as source energy model, instead is just a source of biomass; and will be explained later.

In plants it is clearer, more perceptible: water is essential to them, because it's the very first step in the set of chemicals reactions that make life possible, photosynthesis, or the extracting of energy from the unfolding of water and for it there are at least three things required: light, water, and chlorophyll arranged in order of abundance in the universe. Despite the photosynthesis process in plants is nearly 350 years of being studied, since Lavoisier and others; there are still some mechanisms not fully or at least well understood at molecular level. It has been a long process, because it wasn't something immediate, first it was deducted that gases discovered only recently by them -hydrogen and oxygen – were necessary for the plant to live and were generated by the plant. Hence the word autotrophic (self-nourishing), but animals couldn't do it, and hence the word heterotrophic (nourishing from others). These processes to understand plant's photosynthesis begin with the knowledge of plants cycles and of its physical structure. In 1640, the work of both Johannes (Jan) Baptist von Helmont (1577-1644) and English physician and cleric Stephen Hales showed that plants needed air and water to grow. In 1700, chemists begin to identify individual gases involved in combustion, breathing and photosynthesis. Joseph Priestley (1733-1804) showed that green plants can restore a poorly oxygenated environment in a way to make them capable again to sustain combustion and breathing.

Physician and medic Jan Ingenhousz (1730-1799) inspired by Priestley's research, found little after that only the green parts of plants are able to revitalize damaged air, or better yet, poorly oxygenated – that means they can attract carbon dioxide from air and release oxygen – and this can only be done in presence of sunlight and water. This was the first sign of the role of light in the photosynthetic process. Ingenhousz also discovered that only sunlight – and not the heat it generates – is necessary to photosynthesis.

In the XIX century, research on photosynthesis was centered on the chemical process by which carbon is "fixed" as carbohydrates. By the late 1800, German botanic Julius von Sachs (1832-1897) suggested that starch is a byproduct from carbon dioxide ($O=C=O$). He also claimed in 1865 that, when in presence of light, chlorophyll catalyzes photosynthetic reactions and discovered that chloroplasts were the ones containing chlorophyll. In 1880, German physicist Theodor Wilhelm Engelmann (1843-1909) outlined that the reactions who capture solar (photonic) energy and turn it into chemical energy occurs inside chloroplasts and respond mainly to light's red and blue colors.

It wasn't until the XX century that scientists began to understand the complex biochemistry of photosynthesis. Richard Willstatter acknowledges there are two main types of chlorophyll in Earth plants: blue-green or "a" type, and yellow-green, or "b" type. American biochemist Martin David Kamen, used the oxygen 18 isotope (an isotope is a variation of the same element, with the same atomic number, but different atomic weight, due to the number of neutrons in it -Madder 1990- to trace its chemical "path" along the process. Thus, he confirmed that oxygen created during photosynthesis came only from water molecules present at the time. German biochemist Otto Warburg found that, under appropriate conditions, the efficiency of photosynthetic process can achieve the 100%, which means almost all the sun energy taken by chlorophyll (only from 400 to 700 nm) is turned into chemical energy.

In 1940, the discovery of carbon 14, a radioactive isotope isolated by Kamen, allowed more precise studies of photosynthesis to be done. Using carbon 14, Melvin Calvin was able to trace all the way of the photosynthetic process. In the decade between 1950 and 1960, he confirmed that reactions to light involving chlorophyll immediately capture the sunlight's energy. Afterward, he studied the subsequent "dark" reactions, so called because they happen even without the sunlight presence. He found carbohydrate molecules begin to form in this stage of the process. Working with green algae cells, Calvin interrupted the photosynthetic process at different stages and submerged the algae in an alcohol solution. Then, using a laboratory technique called paper chromatography; he analyzed the cells and the chemicals that had been produced, identifying at least ten middle byproducts created in merely seconds. This series of reactions is now called the Calvin Benson cycle.

Photosynthesis is a key component in a cycle that not only keeps life on Earth possible, but also keeps the carbon dioxide and oxygen in balance. Bruno and Carnegie (2001) emphasized the fact that plants turn carbon dioxide into glucose and oxygen which animals use in a combining process with food, to release energy from it, what is called breathing. Breathing is thus, the reverse process to photosynthesis. During breathing oxygen gets used and there is a generation of carbon dioxide and water, while plants use water to begin photosynthesis. This difference marks a basic distinction to classify the animal and plants kingdom, since, for instance, in animal kingdom there are no known cases of animals containing or expressing chlorophyll.

CHAPTER 2

Photosynthesis in Humans Beings

Arturo Solís-Herrera, Ruth I. Solís-Arias, Paola E. Solís
Arias, Martha P. Solís Arias.

Introduction

If we do not take in account the hitherto unexpected intrinsic capacity of melanin to dissociate and reform the water molecule, then we don't know too much about the origins of life. Life´s generation requires at least four things: The Universe, Light (visible and invisible), Melanin and water. However, their particular natures are beyond our understanding. For instance, the boundaries of the Universe are unknown. We don't know for sure if it's expanding or contracting. The big bang theory has fallen down. The law of universal gravitation doesn't explain the actual shape of the Universe, because, according to the former, the Universe should be imploding, since it misses 500 per cent of extra matter, of mass, given that it will require such mass amount to explain why it hasn't collapsed yet. There are speculations about the existence of additional matter, which, being undetectable, it's called dark matter and in my opinion it is melanin.

Then the sun, this huge gas mass that radiates almost unimaginable amounts of energy from the fusion of hydrogen atoms to conform helium atoms, withholds many secrets. The

Earth is also unknown in many aspects, since the deepest perforations have only achieved to scratch its crust, only 10 km at best. Water also keeps fascinating secrets still. We accept that its traditional formula H_2O, it's only for academic purposes, since the real one remains unknown. It's unusual that it keeps a liquid form when on the outside, given that similar molecules in the same conditions acts like gases. Little is understood about its high boiling temperature, its inclusion complexes, its hydrogen bonds, its ever changing nature, etc.

Also, little it's know about the melanin molecule that has been of general interest since centuries, but, due it's difficult to be studied on lab, has been thought as little more than mud, because it resist the usual methods of research, despite having being studied, even in first world laboratories. The conclusion remains the same; melanin is hard to work with within a lab.

Resuming all the above, the fundamental elements of life largely outgrow our abstraction capacity, since we aren't even able to imagine an explanation for all of them. We don't understand them, but we agree that these elements – and I assert here that melanin is one of them- are necessary to start the dynamic and changing processes that we call life.

Until now, we assume that first was the Universe, then the Earth, next water, a little bit later the ATPase, that is, the enzyme that synthesize the currency of the universal energy exchange molecule, the Adenosine Triphosphate (ATP), next chlorophyll, then life.

This series of events face some contradictions that haven't been resolved. For instance, the ATPase- the suffix "ase" means it doesn't spend ATP to fulfill its task – is an structure formed by amino acids (compounds that has in its structure amines – NH_2^- and acids – $COOH^-$). Nevertheless, it's a very regular and very uniform, built by a number of amino acids (500 amino acids) that aren't randomly disposed, but the opposite. It has such order that nature repeats it over and over again. This order in the Universe or in Earth itself is rather hard to take in just as its been proposed until now, as something a-biotic, that is, molecules brought together by lifeless entities, that somehow are able to synthesize time over time the necessary quantities of such enzymes until, after millions of years the first molecules of chlorophyll were built and slowly afterward the rest of them were too. The former text is akin to pretend that all the complex machinery of a clock fell from the sky, got to Earth in such way that they just were combined, made the clock and, the most amazing thing of all, the clock began to work perfectly.

It's hard to concede such a miracle, because it should be repeated over and over. Even until today, it would be an event so regular in its occurrence that, eventually, the rest of the chemical reactions that compose life are intertwined in its exact current shape and that hasn't change since the dawn of time.

Life is not a random and casual series of chemicals reactions. The expression of life needs that the same exact order which created it keeps occurring. In this sense, it's always the first reaction or number 1, then number 2, the number 3 y so forth, until comes a moment when, due to the high level of amino acids that compose organisms, little but constant differences begin to happen, for example, the possible combinations of number and location are infinite. Suffice to say that the musical notes are only seven, but the number and variety of music created and still being created with them seems endless. If we meditate that the number of amino acids presents in the earth organisms are 20, then the combination possibilities grows exponentially. We could assert that there are more amino acids combinations than melodies. These small differences, that are amplified as the chemical reactions occur, are the ones who produce the origins of the species.

But let's go back to the primary reaction. The question even today is still, in which manner the energetic requirement for the first reaction that generated life was supplied? We understand that energy is defined as the capacity to make a job, which can be of several kinds, something that moves, something that contracts itself, something that expands itself, something that is combined with another thing, etc. On the other hand, NASA defines life as a series of chemical reactions self-sustainable that eventually displays a Darwinian evolution (survival of the fittest).

In regards the first chemical reaction? How did it happen? It couldn't be by chance, because if so, its same randomness would have made almost impossible the generation of life, if it happened once in a while and who knows where, how could the other elements have aligned themselves? It would have been unthinkable that, amidst the chaos life would appear, even if life isn't perfect, it does show a relatively constant order. Nevertheless, when such variations go over a limited range, life just doesn't express itself.

Then, we have to go back to the initial reaction, a primordial one for the origin of life and rephrase the question: How did this energy, in such adequate time, shape, quantity, range and other necessary qualities, generated in such a way as to be able to produce, through millions of years, the series of chemical reactions that compose life itself?

We can write a lot of theories about, just as it have been made so far: an alien origin, divine origin, chance, a-biotic origin, etc.., but if we consider melanin, our understanding changes radically.

Why?

Until now melanin has been neglected; was considered as something of a stigma, and a stigma that has been vanishing as nations advance in its civilization process. 99% of scientists and common people think that as with water, everything is wrote about melanin, there is nothing more to do, to study, to discover. But no, no way at all, both substances harbor fascinating mysteries.

Suddenly, the attention of a growing number of scientists has focused on melanin. There is a very powerful reason for that: en 2002, because of a research started in 1990, on the 3 principal causes of world blindness (Age-related Macular Degeneration, Diabetic Retinopathy, and Glaucoma) due to their incidence and prevalence has not change since fifty years ago or more, which means that the treatments are not working, the intrinsic property of melanin to split and reform the water molecule was unraveled. The research project was planned using mathematical models and digital imagery equipment in order to study the living patient. Our main aim was the development of new therapies for the diseases above cited. However, we found a series of anatomical and therefore biochemical events that led us to experiment and to probe scientifically a sound reality that we can resume as follows:

Melanin is to the animal kingdom as chlorophyll is to the vegetable kingdom.

This concept has provoked and continues to do so, a very meaningful impact, is has the same impact if we put it this way:

Melanin is the human chlorophyll

This is the phrase we coined long years ago to draw attention on an undeniable fact: the existence of what we can call the human photosynthesis.

Photosynthesis in Human Beings

Such as assert hasn't been always well received, for many an occasion we have been the target of jokes and wagers against us from renowned personalities. But that is what our studies on the human retina, as part of a project we started in the Center of Human Photosynthesis Studies in 1990 – it wasn't named like that then; its name was Clinic for People with Eye Diseases – on the three main causes of blindness to find new therapies.

The first step:

The fact about it is that, despite all the technological and scientific advances, the prevalence and the incidence of those diseases remains the same, this means that the recent treatments aren't efficient or effective. Our studies allowed us to, by means of mathematical models, enhance notably the sensibility and the specificity of the methods we used, and to corroborate the rightness of our conclusions. It also allowed us to sense, to sustain and later, to defend the extraordinary importance of melanin in the physiology and pathology of eye diseases. We saw that the human retina is an exquisite sensible to oxygen variations tissue, or, better yet, to its level inside tissues and blood. We were able to watch and record digitally as well as *in vivo*, in a harmless way as part of routine procedure observations, the vascular variations in the retina that come from the structure-properties ratio of melanin inside the human eye.

It took us 12 years to understand the vital role of melanin in the eye as well in the whole body. It wasn't easy, especially if we take into account that laboratory data are so few, ought to melanin being a substance with a extraordinary complex molecular dynamic, way beyond we are even capable to imagine, methods used by researchers so far was akin to take photographs and then, with that sole photograph of, say the Moon or the Sun, try to understand the whole Universe, or at least our Galaxy.

The office where the research was done

Plain and simple, it's not possible to understand the role the molecular dynamics of melanin inside the laboratory. To understand one of the most important properties of melanin, at least the ones related to the origin of life and its role in the concert of biochemical reactions, is imperative to observe the tissue that contains it, watch it live, once and over, trying to discern the best the context that surrounds it, in such way that its actions, its effects and its deficiencies where mounting up to a coherent mental scheme, which finally, after long twelve years suddenly took us by surprise: melanin dissociates the water molecule, and even more; not only breakdown the water molecule, melanin can support the reaction in reverse. The process can schematize in the following way:

$$2H_2O \leftrightarrow 2H_2 + O_2 + 4e^-$$

But what does that mean?

That the hitherto unknown very first spark in life´s origin finally has been found. Let's go back to the first reaction, the primordial one, the beginning of life itself, the one that, according to NASA can be resumed as follows: it had to be a self-sustainable chemical system which eventually entered in a Darwinian evolution.

And why does the dissociation of the water molecule by melanin fills such description? Because, when the water molecule, H_2O, breaks, it dissociates and it unfolds, leaving us eventually with $2H_2 + O_2$ that is, molecular or diatomic hydrogen (H_2) and oxygen (O_2), and to balance the equation we simply need two water molecules dissociating at the same or almost the same time, which stops them being in liquid form, because they are changed into gas phase. Perhaps $2H_2O$ can be considered the minimal formula of water, even more because H_2O doesn´t fit with the most common form of Oxygen, that is O_2 or elemental oxygen, so far, the most abundant one, by other side, O or nascent oxygen is very rare because is a highly reactive form.

And what does that implies?

The answer, in accordance to the scientists' requests, is simple: we have free molecular hydrogen, whose value lays in being the part of water that is going to carry the energy of the broken link. Let's consider the hydrogen as an energy carrier. It's not like any carrier, and it's also not the perfect one, but is the most used by nature, added that to the fact of being the smallest atom (hence the extraordinary difficulties to study it and for instance bottle it up for, let say cars) this same smallness makes it very practical, because it can be nestled on any corner of the eukaryotic cell, which makes it very maneuverable, being an constant, useful and necessary energy that reaches all places inside the cell covering entire cytoplasm and organelles. It's a collective mistake, as happens in every other area of knowledge, to think that oxygen is more valuable than hydrogen. We need only to remember what plants do with oxygen once chlorophyll dissociates the water molecule; they sent it back to the atmosphere, almost as waste or maybe exactly as waste. However, we, mammals, use it a little bit more: we stick electrons to it and send it back to the atmosphere, which makes it a little more usable, because we form with it the most oxidized form of coal: CO_2 (carbon dioxide, $O=C=O$) the well know greenhouse gas.

It's understandable that oxygen were more easily knowledgeable, as its 16 times bigger than hydrogen and hence, way easier to detect than the very much more valuable hydrogen. If we imagine for just a moment those ancient times where life did not existed, only Earth, with water bathed by the sun, with primal lakes and rivers, and if we accept the ever presence of melanin, something that is not hard to imagine if we know that melanin forms even spontaneously, and is stable within water, perhaps during millions of years (a proof of such thing is the detection of melanin molecules floating, traveling in outer space – cosmids it's their name – indistinguishable from neuron-melanin of mammals).

Due to the light-years of distance, above photographs show enormous masses of a dark matter, probably melanin, the most stable compound man ever known (million years). (Source of photographs: http://apod.nasa.gov/apod/ap111025.html)

Going back to those hypothetical primal rivers and lakes and to melanin dissociating the water molecule day and night along thousands and thousands, or even millions and millions of years, making hydrogen carried chemical energy available to its environment, free chemical energy is the only way in which living organisms, both past and present, living in our threatened planet are able to use photonic energy, that is, energy traveling through electromagnetic radiations.

Theretofore, the point I'm trying to reach is that solar energy is the one who pushes life on Earth. It happens every day, as it has to, to allow chemical reactions which eventually conform a system that eventually become self-sustainable along millions of years by trial and error. We can sustain that living organisms can't take solar energy directly, because a transducer is necessary, in other words, something able to transform mostly sun-generated electromagnetic radiations and secondarily from other distant space sources, into chemical energy.

I sustain that life only can be explained through this mechanism, unique since then and till this days: sunlight + melanin + water = free chemical energy, arranged in order of abundance in the universe. In other words, electromagnetic radiations + melanin +contained energy inside the links between hydrogen and oxygen = molecular hydrogen carrying energy (chemical at this point). I'm convinced and I propose that we accept this: the hydrogen and its valuable charge of energy, released from water by melanin, using the energy of the sun, is the explanation for the origin of life on our planet.

The universal photosynthesis system is composed by Light/ Melanin/ Water, especially liquid water (arranged in order of abundance in the universe). And it's the same so far and since the beginning of the times. (Photographs: http://apod.nasa.gov/apod)

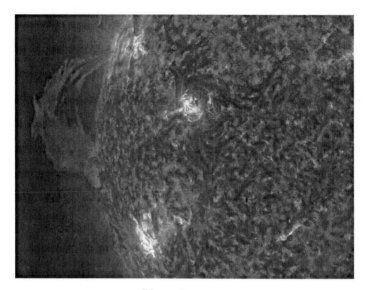

Photonic energy

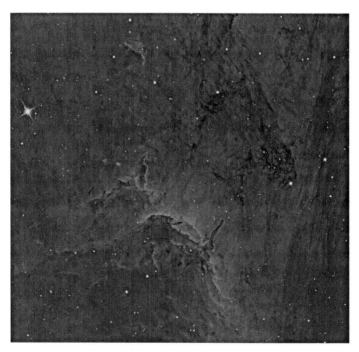

Melanin in the space.

Frozen Water

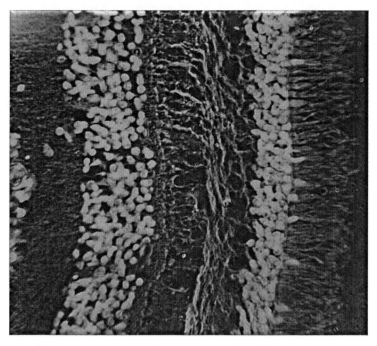

Histological section of human retina, Solís-Herrera, 1998.

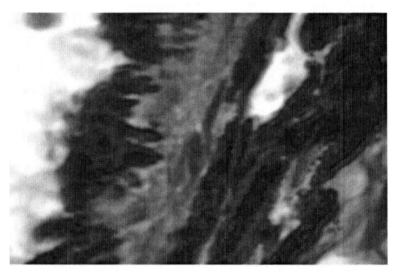

Melanocytes in human choroid layer Solís-Herrera, 2000

CHAPTER 3

For the human body,
Is it light only good to see?

What is commonly known as light or electromagnetic radiations is one of the most important manifestations of energy in the Universe. The latter can be defined as the capacity to do some work, but a definition of what we call light is more complex. Let's just say that we don't even understand its nature, since sometimes it behaves as a wave and sometimes as a particle. In other words we don't fully understand its nature. Let's try an example: if we could amplify a ray of light we could see that it gets to a point were light is indistinguishable from matter itself, because in a certain moment, we wouldn't know if we are looking photons – the elemental particles of electromagnetic radiations – or watching leptons, quarks or gluons, to name a few. To an infinitesimal level, matter and energy are indistinguishable. Light is an extraordinary complex phenomenon, way more that we can imagine.

As light is mostly composed by energy, hence the expression that nature uses it in the whole Universe to carry energy from one place to another. Energy that comes from the sun reaches the Earth trough electromagnetic radiations and the part of it we are able to see we named it visible light. This energy, visible and invisible; is the one who makes life on our planet possible.

We already said that NASA defines life as a self-sustainable chemical system which eventually enters in a Darwinian evolution. So, it had to be an initial chemical reaction starting the following others, since this initial reaction could happen with a necessary continuity, it was only matter of time that a series of chemical reactions happened together, those which, brought together, we called life. But this very first chemical reaction has a cost in energy. This is an energy expense, one that has to be payed. Is sensible to think the energy required to bring forward this first chemical reaction came from light. The history of the Universe, and Earth along it, accords with it. The energy of the sunlight, and the one who comes from the whole Universe, are bathing the Earth day and night since the dawn of time.

Pleiades Deep Field

Stanislav Volskiv

Energy and melanin 400 light years away

But until now researchers and general public thought that the human beings couldn't use directly this luminous energy, given that only chlorophyll, contained in plants, bacteria and algae, was able to do so. Of the five organic pigments known, only chlorophyll possessed, apparently, the power to transform luminous (photonic) energy into free chemical energy, which can be used directly by eukaryotic and prokaryotic cells. The other four pigments - carothenoids, bioflavonoid, hem- pigments and melanin- were not supposed to have this valuable quality and to have other functions, apparently important, but very less transcendental.

About melanin, we could only find in literature that its functions were more akin a sun screen (protection against the damaging effects from solar radiation), for social interaction and protection (mimesis) (Andrezj Slominski 2004). Then, to human beings light serves mainly lighten up some objects as to be able to watch our surroundings, and, with the exception of the synthesis of D vitamin, there is no other relevant role for solar radiation in mammals. In opposition, on plants we accept without arguments that they are not able to survive without water and light, since in absence of this two factors they disappear completely.

Accordingly, the answer to the question posed above would be: yes, light only serves the human body to see well. It was assumed that to use the luminous energy, this has to be transmuted (transformed from one type of energy to another) by the plants first, so we get the final result of it trough the ingestion of food. In other words, based on classic scientific literature, we are unable to directly use the energy contained in light. This can only happen thanks to ingestion and the energy we need diary comes completely from lipids, glucose, proteins, minerals, vitamins and alike from our diet.

However, if we stop to think, we realize that life is not possible without light, whether visible or invisible. For example, there haven't been communities living below the earth. Furthermore, there haven't been any big cities or big malls functioning all time, because humans being tend to get sick and die fast. The initial or more usual manifestation, although not the only one, is alterations of the immune system and depression.

Until now, there was almost a heresy just to suppose the existence of another source of energy for human beings, whatever it could be. It was supposed that human beings, and with them all the animal kingdom, only use light to see. The exception confirming the rule was the tiny part from radiation that we use to synthesize D vitamin. The most solid proof of this is our inability to see at night, depending on the moon phases, thanks to the solar light reflection on the moon surface, for seeing at least partially around us, not nearly as good as when day is, but the darkness of night is diminished, since total darkness is terrifying. Irrational fear of total lack of light is innate.

Our night vision is way behind day vision. The color, brightness, and the definition of objects, plants or animals have no equal.

In a word, light is necessary to see, but just because when it reflects on things, human retina can grasp forms, movements, colors, etc. We have thought it's only because of this and this only, never thinking the energy carried by light beams could be used in meaningful ways by human tissues, namely retina or any other.

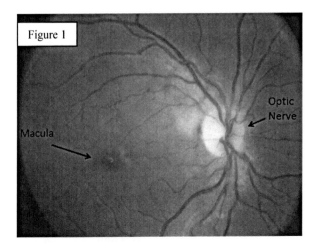

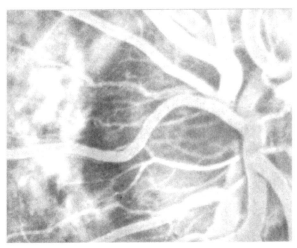

Figure 1. Photograph of the human retina, right eye. The semicircular structure on the right side, from where blood vessels (red lines) are coming is the optical nerve, whose dimensions are approximately 1200 microns, roughly the equivalent of 12 human hairs, which average at 100 microns each.

CHAPTER 4

The Photosynthesis Process in Plants, the Role of Chlorophyll.

Photosynthesis, from the Greek Photo (light) and Synthesis (putting together), can be schematizing as follows:

$$6CO_2 + 6\,H_2O \rightarrow C_6H_{12}O_6 + 6O_2$$

Plants kingdom and the animal kingdom have substantial differences and one of them is that plants are able to convert photonic energy into chemical energy and the animal kingdom not at all. Or at least, that was the extended belief. Photosynthesis has a transcendental and necessary first step, the dissociating, partition or unfolding of the water molecule, which allow us to obtain hydrogen and oxygen. This is so important that plants life is not possible without it, it simple doesn't express.

Photosynthesis in Plants.

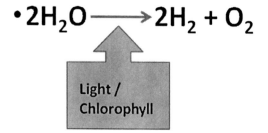

$$\bullet 2H_2O \longrightarrow 2H_2 + O_2$$

Light / Chlorophyll

Chlorophyll molecule, where we can observe the reaction center, formed by 4N (nitrogen).

chlorophyll a $R_1 = CH_3$
chlorophyll b $R_1 = CHO$

Chemical Formula: $C_{55}H_{71}MgN_4O_5^{2-}$
Exact Mass: 891.53
Molecular Weight: 892.48
m/z: 445.76 (100.0%), 446.26 (73.2%), 446.76 (41.1%), 447.26 (11.2%), 447.27 (4.2%), 447.76 (2.9%), 446.26 (1.5%), 447.77 (1.2%)
Elemental Analysis: C, 74.02; H, 8.02; Mg, 2.72; N, 6.28; O, 8.96

But, why is it so relevant?

There are several reasons for it:

1. It happens at environmental temperature and this seems so natural and so simple because we observe it happening everyday among all plants. Proof of this is the blooming and fruits from plants, apparently without anything else but sun´s light and water. But there is a catch: if we want to reproduce the split of the water molecule, to obtain diatomic oxygen and hydrogen as plants do, we need to

ARTURO SOLÍS HERRERA, MD, PHD.

heat up the temperature to 2000 °C since at 100 °C we only get water molecules as steam, but hydrogen and oxygen are still together.

2. The main component resulting from this reaction for plants is diatomic hydrogen, because of the simple reason of it being an energy carrier, the cleanest form of energy known. The vegetable cell is going to use this valuable hydrogen to energize the following reactions that lead to the synthesis of, let's say, glucose ($C_6H_{12}O_6$). We must remember that plants absorb carbon dioxide (CO_2) and, as a result of photosynthesis, this molecule, the most oxidized form of coal, the plant cell adds mainly hydrogen, because for each carbon dioxide molecule plants add to it 12 hydrogen and 4 oxygen atoms to form the main source of carbon or biomass in both animal and plant kingdom, glucose (glucose 6 phosphate, in biochemical terms, when the organism adds a phosphate to it).

Chemical Formula: $C_6H_{12}O_6$
Exact Mass: 180.06
Molecular Weight: 180.16
m/z: 180.06 (100.0%), 181.07 (6.9%), 182.07 (1.4%)
Elemental Analysis: C, 40.00; H, 6.71; O, 53.29

Glucose

52

Amusingly, plants seem to dispose the diatomic oxygen, a very stable molecule. In fact, plants send it back to the atmosphere without using it at all or few molecules at most, unlike hydrogen, which is exhaustive and widely used by the cell, because energy is released similarly by chlorophyll and melanin, through diatomic hydrogen, $2H_2$; but energetic efficiency of chlorophyll is thousands times less than melanin; but in both cases, energy is released symmetrically in all directions. It is a relevant fact that diatomic hydrogen doesn't combine with water; so cytosol is bathed entire and continuously by growing sphere-like of diatomic hydrogen whose starting point is the chlorophyll in plants and melanin in mammals ; therefore molecular hydrogen is not only strongly treasured by NAD and FAD, molecules who take it and works as reserves for it, because they give it or take as the several chemical reactions, happening constantly inside the cell require, and which gives as a result what we call life. NAD turns into NADH (reduced) and FAD turns into FADH (reduced) once they capture hydrogen and its highly valuable energetic charge. NAD and FAD are identical in plants and humans.

Until here, we find two similar situations among plants and animals. First, the human body seems to dispose the oxygen, although not as fast as plants, for instance it sticks electrons to each oxygen atom and also sends it back to the atmosphere as carbon dioxide (the most oxidized form of coal O=C=O) and, secondly, human beings and mammals also have NAD and FAD, molecules with identical functions in both kingdoms, to capture hydrogen.

The important thing in this chapter is to understand the first step in photosynthesis, the unfolding, partition or split of the water molecule. The energy the cell plant requires for it comes from red and blue light. In capturing the light energy through chlorophyll and using it to unfold or break the water molecule, the transformation of light energy into chemical energy is happening, the plant is able to use directly this luminous energy, even if it's restricted to a wave length between 400 nm and 700 nm. The rest of the visible spectrum is reflected, that's why we see plants green, even if it's a known fact that greens do not exist in nature.

The absorption of light energy by a pigment and, as a result of it, the beginning of an ionic event is a common phenomenon in nature, but until now we accepted it only in the plant kingdom, cianobacteria and algae, that is, all those living beings possessing chlorophyll.

There are examples of melanin in the plant kingdom – eggplant is one of them – but examples of chlorophyll in animal kingdom seem to be rare. One recently observed happens in the butterfly, but only in the caterpillar phase.

Eggplant samples with melanin and albino (low amount of melanin).

If plants "breath" carbon dioxide (CO_2) where did the CO_2 come from if we know that plants came first and after them the animals? Yet, animals seem that are the main source of CO_2. This already speaks confusion if we stay stubbornly on classist theories about the origin of life.

CHAPTER 5

Water is indispensable for life but, what are its functions within the human body?

The next definition, taken from a medical dictionary goes like this: "Water (H_2O) Chemical compound whose molecule is constituted for an atom of oxygen and two of hydrogen. Almost three quarters of the earth surface are covered with water, which is essential for life and forms more than 70% of living matter (it represents 65% of total human body weight). Pure water freezes at 0°C and boils at 100°C at sea level". Do take note that the author uses the phrase: essential for life.

Yet another definition, this one from a chemical dictionary, it says: "As a polar liquid, water is the most powerful solvent known". This is the main role water has been granted, a solvent. This means that water was important because is the perfect matrix to the chemical reactions that compose what we call life. In other words, it was considered as a marvelous substrate, a kind of support, solvent at most; with few or nothing to do with the generation of energy, except on plants. From where comes from the energy to animals? It was thought that all energy came from food, even if some cell biology books accepted that only 90% of the energy required everyday came from food, not specifying where the missing 10% came from did.

The role of water in biological processes

Life's chemical and physical processes require molecules who are able to move, find each other and change links (covalent means sharing electrons, although there are other kinds of links as Hydrogen bonds and Van der Waals forces) this represents a very dynamic process, always changing, ceaseless inside the still not very well known, really complex and complicated metabolism's and cellular synthesis processes developed by nature along four billion years of evolution and therefore beyond our understanding, even so, they are processes indispensable for the cell to express life. On the other side, a liquid environment, allows molecular mobility, a very appropriate vehicle, also a good heat control trough sweat (for instance) and water is not only Earth's most abundant liquid, but also is notably well fit, for this and others purposes, in the well functioning of the organism, although never before was suspected that humans and/or mammals were able to uses it as source of electrons and/or diatomic hydrogen. Let's examine with some detail the physical and chemical properties of water.

Structure, activity and properties of water

Even if we tend to take water properties for granted, actually it's a really complex substance. If we compare water with other compounds of similar molecular weight or similar stoichiometric characteristics, we find several notable things. Most of these compounds (CH_4, NH_3, H_2S) with a low molecular weight, are gases at room temperature and have boiling points way below water. We have to remember that one of the chief properties of water is to be considered the universal solvent. Thus, a gas couldn't work physiologically as water does. This important property of water to act as a solvent, as support for the other chemical reactions that help the right functioning of the body take place, is explained by the tendency of water to form hydrogen bonds.

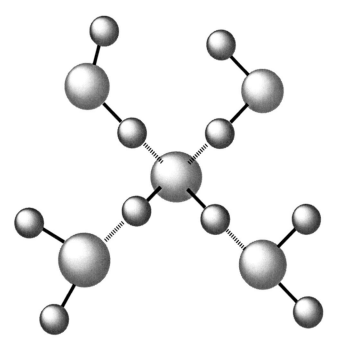

Figure 3. Example of hydrogen bonds.

Each water molecule is at the same time a giver and a taker of hydrogen links. As a consequence, evaporation of water needs an exceptional amount of energy for a molecule of its size. Hence, both the evaporation heat as the boiling point are exceptionally high on water, especially if we compare it with molecules of similar stoichiometric characteristics as noted above (CH_4, NH_3, H_2S). Water, finally, stays liquid at the weather conditions of the most of Earth surface.

When water freezes, the hydrogen links between molecules became more regular and better defined, shaping a rigid tetrahedral net in which each molecule is linked to other four by hydrogen bonds. This net structure falls down only in part when ice melts and remains with a certain order for a long time, even at higher temperatures. Liquid water structure has been described as blinking groups of hydrogen bonds with traces of the ice net that breaks and form again as the molecules continuously moves. The relatively open structure of the ice net explains another of the uncommon properties of water: in liquid state it has more density than when solid, because when the net breaks, molecules can come closer. This fact apparently trivial is of the utmost importance for life on Earth. If water would behave as the majority of substances, whose density is bigger when frozen, the ice forming every winter in lakes and oceans would drop to the bottom. Once there, isolated by the covering layers, it would be piling up with the years and then most of the water on Earth would be contained as ice.

Comparing with other organic liquids, water has more viscosity, as a consequence of the stuck structure from the hydrogen bonds. This cohesion also helps to explain the high superficial tension of water. The high dielectric constant of water comes from its bipolar characteristics. An electric field generated between two ions produces a notable orientation of the water dipoles on it, as well as a meaningful amount of induced polarization. These orientated dipoles make a counter-camp, lowering the effective electric force between the two ions.

In physiology books we can find the following: "we take water when thirsty, we also take food to obtain energy, but food always has large quantities of water". Water molecule, as solvent; is used as a vehicle to expel the normal waste that result from body metabolism, for instance to hydrate ions and to control body heat trough sweat. Also it helps us to keep a blood level stable, so its volume affects arterial pressure. It's also the main ingredient of interstitial liquid and of inner cell liquid, but is never mentioned the paramount importance of extracting energy throughout it, something fundamental to life, because without a reliable source of energy, the cell dye in seconds, therefore the energy that cell needs is taken trough the photo- system composed by Light/ Melanin / Water; arranged in order of abundance in nature and by analogy with plants called human photosynthesis. These processes represent 99% from the total we need every day, leaving only to meals or glucose the role as source of carbon chains or biomass.

Intrinsically speaking, this doesn't resolve the doubt about that the physiology books tell us, why it is indispensable for life? According to the written here above, water in the human body is only good to carry substances, inside the blood stream, in urine, in sweat, inside the intestine, in the cerebrospinal fluid to name a few examples. But, at the end, it seems hard to accept that a substance that only seems to support, to uphold, to carry, to dissolve, or cleaner at most; could be so indispensable for life. In other words, without water there's no life, without water, life cannot express itself and this works both for plants and mammals, the answer is that water and only water with its strange and unique properties, perfectly fit with melanin so the later is able to express its intrinsic capacity to split and reform the water molecule.

How is it possible that a human being would die relatively rather fast, in a few days, maybe three (depending on the environment, light, temperature, pressure) without ingesting water? Doubt comes when we keep in mind that water is a solvent, perhaps the most well known and out of discussion most used by Mother Nature, turns out is not the only one, and that it could be life based on other solvents, like methane and ammonia. Maybe this couldn't happen on Earth because, at room temperature, they exists more as gases than as liquids, but in other planets with adequate pressure and temperature condition, it could be possible to find such phenomenon, that is, life expressing itself without water. Hence the description of beings very different from us in science-fiction stories about life in other planets, of which hasn't been found until now even the most remote sign, because, accordingly to our present knowledge, without water, there is no life at all, and we could add, without Light and Melanin either.

Life as we know it doesn't express itself if there are no sunlight or electromagnetic radiations, melanin and water. This works for both plants and animal kingdoms, because plants cannot survive without water also. In plants, water functions are very similar, given that water works as a solvent, as diluents, as volume, as heat control through evaporation, and other known and unknown functions so far these functions are like the ones in human body. However, the main role of water in plants, the generation of energy from sunlight, what is called photosynthesis, it wasn't supposed to happen in humans. Also it wasn't supposed to exist in animals the generation of energy throughout water dissociation and reformation. It was only accepted (until now) to be happening on plants, cianobacteria and algae, but to the contrary, in humans, in mammals, there didn't seem to be any example of it in the animal kingdom at least apparently.

CHAPTER 6

Melanin: the human chlorophyll.

As we mentioned in the former chapter, melanin has only been assigned functions as a protector against sun radiation (as a sun screen), for mimesis and for social interaction. Outside that, it doesn't seem to be anything more. The apparent stability of melanin has been precluding biological functions apparently relevant. However, if we stop for a moment and search for more information about it, melanin is a compound that has marveled (and keeps doing it) plenty of researchers and scholars during centuries. Let's propose an example: we have theoretical formula for melanin in the next image:

Theoretical formula of melanin

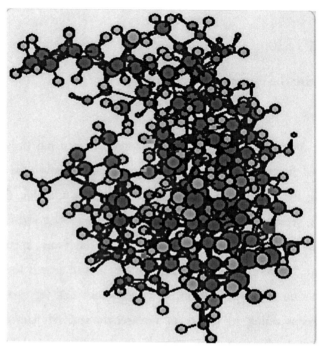

Three-dimensional theoretical formulae of melanin

Let's try to explain it in another way. We see plants as green because chlorophyll only absorbs light in the range of 400 nm (blue) to 700 nm (red). The rest of wave lengths are reflected, and that's why we see them green. Opposite to this, melanin doesn't reflect or emit any wave length; it captures absolutely all of them. Hence is the explanation of why we see the melanin black, which means, dark. This is the main obstacle, the one that makes melanin a mystery. Einstein already said it so "we can study light, we cannot study darkness" and melanin is the darkest molecule man ever known.

We must see that melanin is not considered a protein, because it doesn't has peptic links. Its structure is fairly symmetric, with the addition that the links related to it brings a notable stability. It is considered that the peripheral portions of the structure are the ones in charge of "cropping" the photons, which starts a cascade of electrons towards the reaction centers (4N), where occurs a potential difference, enough to attract the hydrogen atom stronger towards the reaction centers of the melanin, instead of the oxygen atom.

Its molecular weight is estimated in millions of Daltons and there are hundreds of reaction centers, represented by the 4N groups in each molecule gram. This is another big difference, because it seems like it was many "chlorophylls" put together. However, many reaction centers means a great efficiency to collect or absorb the energy from the electromagnetic radiations and that means a greater number of water molecules unfolded in an amazingly short time span, because it takes to collect the energy to break the molecule approximately $3X10^{-13}$ seconds.

Why isn't known, or better yet, why we cannot know the formula of melanin?

It's due to the same reason: because it absorbs all kinds of energy and doesn´t gives out anything. It's something known that melanin captures all wavelengths from the electromagnetic range, this means it absorbs the full known electromagnetic continuum, not only the visible part of it, which is a rather small part, that goes only from the 400 to 700 nm wave length, instead the absorption spectrum of melanin goes from radio waves (100 Mm) to gamma rays (1 pm).

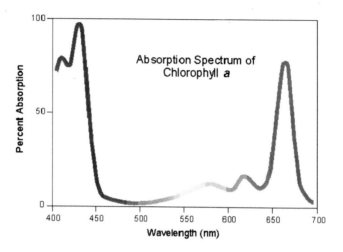

The absorption range of chlorophyll (above image) where we can watch the absorption pikes around the 400 to 700 nm; compared with the absorption range of melanin below: that is the whole electromagnetic range.

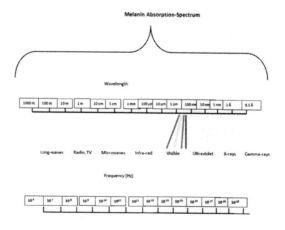

Thereby, melanin is the darkest substance man ever known. Available laboratory research of today are based on the application of some form of energy to the substrate or substance been researched. Such applied energy usually gets absorbed by the studied molecule, which transforms it or processes it according to its structure and sends the energy back, in compliance with the law of conservation of energy. However, the energy emitted by the number of known compounds has enough differences among them, which helps to discern subtle differences that allow afterward the identifying of its composition or nature.

And here lies the problem, melanin absorbs all wavelengths, it's been said that such energy disperses in a non-radiant way, and because of that, it doesn't sends away any signal.

As a response to the energy applied and absorbed, whatever it may be the nature of the energy applied, melanin consistently absorbs it all and doesn't seems to be given anything back in return, something that goes against every law of energy conservation, but we are aware that is not the same thing that it break the laws and another that we are not able to detect the resulting energy.

Dr. Helene Z. Hill, researcher, published in 1982 establishing that with melanin occurred something akin to what happened to the six people who were looking at an elephant. Nobody knows melanin as a whole. We only are able to watch small pieces, tiny parts of a huge puzzle, some small properties, but the immensity of melanin it's so uncommon that goes way beyond our reflection capacity, even beyond what we are able to imagine.

Then, let's take my proposal for good: melanin uses the energy it absorbs to break the water molecule, obviously, in presence of water and light, both visible and invisible.

This happens every time the required conditions are met, for instance, in humans as well as in plants, the process of getting energy from water is a basic process for life to express itself. Same as happens with plants and chlorophyll, in humans melanin is the compound that allows the series of reactions that leads to life begin. When we are able to understand the water´ dissociating by melanin, in the presence of photonic energy (the one being carried by electromagnetic radiations), then the phrase of water is being indispensable for life gets more meaningful. Otherwise it would seem contrived that a liquid only good to sweat, urinate, softening organic waste, use space, as inner cleaner, to generate volume and for the dilution of compounds, were so indispensable, so irreplaceable. In fact, when we are searching for life other worlds, the first thing to find out is the presence or absence of water (perhaps in the future, there will also be a search for the presence of melanin).

If we include in our mental picture the energy we get from water, and that we get it thanks to melanin and electromagnetic radiations, it seems simple to accept it as chemical reaction that apparently has no cost to us, since the energy required to break the water molecule with the photonic energy obtained from electromagnetic radiations. Thus, things look different now. Inside the eye, 99% of the daily required energy will come from light and water, and if we take this ratio to the whole body, then we have the major part of energy required coming from electromagnetic reactions and the following dissociating of water by melanin. This means that there's a very important energy source that has been overlooked, since it comes from a previously unthinkable place, since before us, there hasn't been a more detailed description of the human photosynthesis.

All this goes against all the other books that claims we get our energy from food, otherwise we doesn't get the precious inner force needed to live. There's truth in this, but not absolutely so, since the triad of light, melanin and water gives us nothing less than 99% of all the energy. The important thing is we are not talking about some marginal or extra energy. It's nonetheless the primordial energy, the starting energy, in other words the very first spark of life.

We have to remember this is the force which brought life itself. This is what allowed that those subtle initial biochemical reactions were self-sustainable enough to give time for the rest of the body's multiple systems we know today to hatch. But initially, deontological speaking, that 99% was everything, and, after that, the biomass body systems were gradually hinged and therefore developing slowly along four billion years of evolution. That allowed the collection and then to metabolize (use) all the other nutrients we learned to recognize in the form of food, hatched; the food we ingest, which slowly became to be the biomass. And the highly complex interplay between biomass and energy is a set of processes beyond our understanding, however now we have identified the very first step, the water dissociation and reformation thereof; and the players: Light/ melanin/ water, arranged in order of abundance in the universe. Anything else became later.

It's important to not forget that in the very beginning there was only water, melanin and light as the source for energy. That was all our initial force. This is what can be called self-sustainable. This 99% makes possible that all the other mechanisms required to completely conform our phenotype hatched with all its wonders, for we don't have anything trivial, useless or confuse. In other words, this 99% is the base for the right functioning of our body and this is repeating over and over again. It was so since the beginning and it will continue to be, a basic 99%, energy that, if wasted, will makes us sick, because it will generate an lack of balance, affecting sooner or later all the other organic functions. We can relate it to a house's foundations, which need to have some required elements, since without them; the house will be built upon sand, paraphrasing the Bible.

If this initial reaction doesn't occur, or happens badly – that is, it occurs outside the needed or adequate limits - the consequent chemical reactions doesn't occur at all or at least adequately, or do it badly, the latter are the ones that allow, for instance, the optimum use of the energy taken from food.

We say for instance, because the full impact of this 99% goes beyond in the context of normal cell biochemistry that it would seem, because there is a rigid order in the using of the diverse energies available (both luminous and from food) and this order it's the same since the beginning of life, because life hatched from luminous energy.

Much later, all the needed chemical reactions were added, as the anatomic structures that supported it so life could be possible. That's why the order remains the same and will be so forever: first comes the harvesting of luminous energy, and immediately the use of the energy contained in food. Anything that upsets this order, o causes lesser or bigger readjustments on it will have repercussions in the cell functioning, as in tissues, organs and systems of the human body, with the occurrence of one or a number of ailments we all know. However, it would be more accurate say sick people than sickness, because every one of us will show these consequences in a particular fashion, the same way everyone is happy in his or her particular fashion.

There are some interesting observations about this. For instance, photosynthesis stops, although not completely, in sickness as sepsis, often defined as the hardest to understand and most difficult to treat of the inflammatory diseases. Another notable example is the one we observe in some species when photosynthesis slows with cold climate, when the animal goes to hibernation. Why it doesn't happen in humans? The answer to this question could be a breakthrough that could stop the impoverishment of the body in prostrated patients, because at the first week of staying in bed, the patient begins to lose approximately 1% of his body mass and bones by day. Lastly, when photosynthesis stops completely, death comes in an amazingly short time (minutes).

At the end, all the energy we take, we need, comes from the electromagnetic radiations. Our organs, our systems can only generate 99% of such energy in a direct way (human photosynthesis) the building blocks that constitute the biomass comes from food.

However, from this point of view, energy directly generated actually constitutes 100%, because the food we eat also comes from electromagnetic radiations and water, but need to be processed, transformed or transducer by different organisms, in this case by plants, algae or cianobacteria (autotrophies).

MELANIN IS PRESENT IN ALL
LIVING FORMS

It is a very interesting fact that photosynthesis has significant effects on the eukaryotic cell oxidize-reduction ability. The experiments shows that when we stimulate melanin's photosynthetic effects, its extraordinary ability to unfold water and thus increase inside the cell the availability of the highly valuable energy carrier which is hydrogen, this redox ability increases fourfold or even more, depending on the time we hold the stimulation and on how do we do it. It's because of this that when we treat diseases like Alzheimer, arthritis, colitis and others, the improvement increases while the patient stays on treatment, this means the physiological and/or therapeutic doesn't have a limit. Besides, when we do this stimulation, the eukaryotic cell becomes more selective on capturing the surrounding environment nutrients, it becomes more exigent, more careful, so more diseases can be prevented when we stimulate photosynthesis as prophylaxis.

Another really curious fact is related to sleep. Until now, nobody knows why all mammals have the necessity to sleep. However, if we can differ between day and night environment energy or on available irradiated energy, it seems sensible to claim that, as life originated with this rhythm, when there is less energy available, in other words less hydrogen and thus less oxide - reductive ability on the cell, then by night our skills decreases notably because the cell doesn't have all the oxidize- reductive ability its able to.

The cell's oxide-reductive ability is a basic element for life. We could say in lives equation melanin came first, thanks to its ability to transform visible and invisible light energy into chemical energy. Then, there was a first oxide-reductive chemical reaction present, because the water unfolding is an example of oxide-reduction and the following reactions, which were slowly linking to each other to express what we know as life, are nothing but more examples of oxide-reduction reactions. One more proof is vegetables also supporting its life form based on this kind of chemical reaction, and for it to happen, flowing energy cycles are required, for, interestingly enough, when we grow plants on controlled conditions, as to be constantly irradiated, the plant obviously grows faster, but with a very unpleasant flavor. Hence it's not a popular practice.

Oxide-reduction reactions can be described as matter exchange between two substances and could be molecules or cells.

They will always occur simultaneously, for a molecule to lose a part of itself, electrons for instance, another substance hast to win it. They don't occur isolated from each others. This matter exchange can by, on the electron level, of one or many, or rather of complex atoms and so on, because they depend on several factors, such as its surrounding, the reactant substances concentrations, pressure and light's amount.

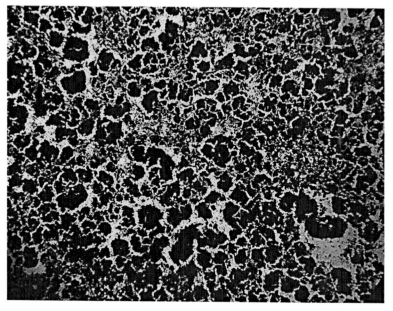

Microscopic photography of melanin, 100 X.
® Arturo Solís Herrera, MD, PhD.

CHAPTER 7

Is it Melanin, the dark matter?

Melanin in our body has a main function besides the energy production: the balance. And the dark matter in the space makes something similar. For instance, the **black holes,** that have surprising effects on the "health" of the galaxy they occupy, and supposedly if there is too little black hole activity and a galaxy might produce a surfeit of youthful stars exploding as supernovae. Too much, and it would suffer from reduced star formation. The cosmic balance can be produced in the same way than into our body and by the same body substance: melanin. a very important fact is that Melanin is the most stable substance man ever known, its half life is million years, and its nature is massive, thereby the molecular weight of melanin is estimated in millions of Daltons.

Dark matter, black holes and melanin have a unique nature. Black holes are very efficient at converting matter into energy, and vice-versa. Black holes can act as gigantic spinning electrical battery capable of expelling matter at nearly light speed across tens of thousands of light-years. The connections between black holes and life are complex, but our galaxy's central black hole seems to have made numerous contributions to our ability to exist at this place and time[1]. Our moderately active black hole strikes a balance, not too hot, not too cold. The connection between the phenomenon of life and the size and activity of super massive black hole is quite simple. Too much black hole activity, and there would be little new star formation, and the production of heavy elements would cease to occur. Too little black hole activity, and environments might be overly full of young and exploding stars or too little stirred up to produce anything. We owe so much to them.

The Melanin inside the body, on Earth and in Cosmic Space tends to be close of energy sources.

In the Space, Melanin adopts different configurations.

CHAPTER 8

Why is the energy we get from light so important?

Before there wasn't even a concept for the existence of human photosynthesis, given the fact that traditional books told us that only plants has the mechanisms necessary to use sunlight to obtain energy directly from water.

But let's analyze this problem. If we go back to NASA's definition about life, it says life is a self-sustainable chemical system, and if we accept melanin is an extraordinary molecule, capable among other things to transform photonic energy into chemical energy, then our concepts about life's origin are radically transformed.

Until now, it was thought the first living beings were plants, and latter animals hatched, among them man, the jump between plants and animals, one that I would call a jump to the void, for plants cannot jump and there are not solid elements to support this idea. Here it is the first trouble with it: with this theory we are accepting the first organic molecule ever was chlorophyll, since, according to NASA, life has to be a self-sustainable chemical system (producing somehow its own energy) and the initial chemical reaction, the first change, consistently enough to allow the later escalating of the thousands and thousands of chemical reactions that together builds what we call life, happened thanks to chlorophyll. However, this line of thinking has some incongruous things that are very hard to explain, given the fact that chlorophyll is an extraordinary hard to synthesize molecule, even today. The odds of that synthesis occurring by chance are null, because with a molecule able to capture sunlight and to transform it into energy to impulse or energize the subsequent chemical reactions, the origin and later evolution of life becomes an unsolvable puzzle.

In short: no even nature, not even by chance, can synthesize the molecule called chlorophyll randomly from zero, or from the primary elements that existed in those early times, the environmental conditions we suppose primitive Earth had. Remember, we take as an axiom (a truth that doesn't need demonstration), a wrong one from my point of view, that the first chlorophyll molecules were created abiotically, a word that can be interpreted as "without relation to life", or else, as something not alive.

But this theory sounds crazy to me for the following reasons: we know chlorophyll only last 20 seconds outside the leaf, unfolding the water molecule and immediately deactivates itself permanently. On the other side, if we concede to the fact that the first or initial synthesis of chlorophyll was because something a biotic, what could it be this famous a biotic "entity"? How could something without life be able to synthesize something so complex, so extraordinary hard even today as a chlorophyll molecule every 20 seconds?

And all here above without taken notice from the trouble of getting this imaginary a-biotic entity the necessary elements to synthesize it, because it is also hard to accept that this imaginary entity, "casually" had everything at hand to create chlorophyll.

However, the problem remains the same; where this imaginary entity took the energy to synthesize that first chlorophyll molecule? And then we must have the three leaf ready to put (chlorophyll) in it, because, otherwise it cannot be that, without leaves, chlorophyll could last those thousands or millions of years for the other reactions came together. I mean, we have only 20 seconds to put chlorophyll inside the leaves of different threes, who implemented or built that previous three leaf? Using which energy? From where and how was the necessary energy to build even a single three leaf took? It would as much as if we were to say that first it was the car, then gasoline or any other nonsense, because, among other things, a-biotic also implies "without voluntary movement", regardless if anybody ever have seen a walking three.

Another argument that shows the theory of chlorophyll being the first molecule not holding water is the little detail of chlorophyll needing to be in a critical point to be able to dissociate the water molecule. Today, is still a dream to be able to hold this molecule in that extremely critic state inside a laboratory, because the required conditions are really demanding and hence, very hard to reach within an adequate standard.

It's almost unthinkable that such conditions were present on Earth for enough time, or at least the adequate time to allow an optimum evolution degree in time and shape that allowed the evolution of species, even of plants they.

Another huge problem about the a-biotic origin of chlorophyll lies in that the experimental works to select a molecule with the required complexity to absorb only from 400 to 700 nm, whether made by nature or the fabled a-biotic entity, also required energy to be made. Then, if we accept that the really complex chlorophyll came from chance, then we need this lottery to repeat itself every 20 seconds to give the other elements or chemical reactions which together form life time enough to combine. It sounds more like a cartoon movie than a scientific theory about the origin of life.

Regardless of these former problems, impossible to resolve as they are, we have yet another one: chlorophyll only produces hydrogen and oxygen from water. This is also a powerful obstacle for the generation of life, because pure oxygen (100%) isn't compatible with life itself, because such high oxygen levels would denaturalized almost anything, or perhaps everything, even more so the delicate and quantum structures of the cell, be it prokaryotic or eukaryotic.

Chlorophyll's wonderful power to dissociate water only happens inside the leaf, because once extracted the molecule, it keeps on unfolding water for about 20 seconds and then deactivates irreversibly. This has not been changed despite the efforts of a number of researchers and renowned universities around the world, who has been trying for at least 60 years without a remarkable success.

It seems logical to think in the use of chlorophyll to extract hydrogen from water to apply it in several things, for instance in energy devices, like a fuel cell, that through tiny ducts on high-tech materials, using platinum as a catalyst, will reunite it again (hydrogen and oxygen) to give us water and electricity.

However, the latter hasn't still be possible, because chlorophyll molecule must be frozen around -20°C to keep its valuable skill, photolytic we may call it, a little more time. This implies using energy that is not going to be recovered, or at least not to allow a reasonable cost-benefit ratio, besides, when taken back to room temperature, it only last for a few seconds before it will be again inactivated after 20 seconds. In other words, there is no time at all. To make it worse, until today, chlorophyll has not been synthesized, yet another huge problem, because imitating the chloroplast isn't something easy at all. There is no technique able to repeat this skill (among many others), something nature has done daily over time.

Plant's photosynthetic mechanisms, began to be studied since 1930 by C.B. Van Niel – a beginner of detailed photosynthetic studies – are "cheaper" energetically speaking, to dissociate or break the water molecule, whose hydrogen links needs a lot less energy to dissociate, comparing with the energy required to separate double covalent links of oxygen and carbon, these ones are way stronger ($O=C=O$).

Luminous energy cannot be used directly to oxidize water and reduce CO_2, what we call the dark phase of photosynthesis. Even more, it doesn't happen in any other known circumstance to this day. This global process we just described is actually separated (in luminous and dark phase) chemically as well as physically, in two sub-processes in each and every one of photosynthetic organisms (or autotrophic which means they are capable of generating their own nutrients).

In the first of these sub-processes, on a series of reactions or steps called luminous reactions or phases, sunlight energy is used to oxidize photo-chemically water (this means, it needs to be done with photonic energy, otherwise, it doesn't happen). With this water oxidation (dissociation) two things are achieved: first, the oxidizing agent $NADP^-$ gets reduced to NADPH, producing equivalent reducers and releasing O_2. Secondly, a part of the luminous energy gets caught through phosphorylation of ADP to produce ATP. This process is called phosphorylation. However in accordance with our research there two mistakes in this process: when ADP is upgraded to ATP the energy is released, and by other side, when ATP is broken down to ADP, the energy is absorbed.

In the second sub-process, the so called dark reactions of photosynthesis, NADPH and ATP produced by the luminous reactions are used for the reducing synthesis of carbon hydrates from CO_2 and water. These reactions were initially called dark to remark they don't need the direct intervention of sunlight. Let's remember all superior plants and algae, photosynthetic processes are strictly placed in the organelles called chloroplasts. But our research shows that the world strictly needs to be changed.

In consequence, what we called photosynthesis only refers to the unfolding or dissociating of the water molecule, using photonic energy as power source. Thus, the similarities between chlorophyll and melanin are well outlined: they are both transducers, which means, they both transforms electromagnetic radiations into chemical energy. Interestingly, or quite logically from a deontological point of view, the next step is rather similar: both processes use $NADP^+$ (both animals and plants use the same molecule) to treasure, so to speak, the highly valuable hydrogen, specially the energy it carries and it can be used by the eukaryotic cell, mainly to impulse or energize the numerous and highly complex biochemical reactions that leads to what we know as life. Here stops the description of photosynthesis, since an exhaustive revision it is not the goal of this book.

CHAPTER 9

Light, Melanin and Water, the Universal Photosynthesis System

Before the evolution of photosynthetic organisms (plans, cianobacteria and algae), Earth's atmosphere, most likely lacked oxygen (this is merely a guessing, not entirely confirmed). It is the most abundant element in the Earth's crust (almost 50% by mass), forms about 21% by volume of the atmosphere, and is present in combined form in water and many other substances. Diatomic Oxygen beside to be the most abundant form of oxygen, circa 96 %; is a by-product of photosynthesis and the basis for respiration in plants and animals.

Pre-photosynthetic organisms must use high energy molecules, synthesized abiotically. Its little more than impossible to synthesize chlorophyll abiotically, that is, outside a living organism, but melanin do can be form easily on past, present and future Earth's conditions, for its presence has been described even in outer space (cosmids). This is congruent, because what eventually gave origin to life must be something able to repeat itself often enough, so if we start from chlorophyll, the chances needed for life to be originated by it became more and more unlikely.

Today's books teach that, without vegetable photosynthesis, this a-biotic energy sources would have been complete depleted and life would have disappeared or not hatched at all.

Known fossils suggest the first living photosynthetic organisms (plants, algae and cianobacteria) appeared at approximately 3,500 million years ago. The slow conversion of the (theoretically) not oxidative primal atmosphere into an oxidative one opened the path for animal evolution. Today is widely accepted that plant photosynthesis constitutes the energy source for almost all living forms. However photosynthesis in plants and humans has some other common characteristics that can change this widely accepted concept.

Things seem very congruent till here, but let's remember chlorophyll synthesis is little more than impossible to be made abiotically. Even today it's not been possible to do, and, furthermore, there aren't theories that explain in a reasonable way how could such a complex molecule been made by chance, and be so all the many times needed. On the other hand, is all but impossible that the required amount of chlorophyll, both in quantity and quality, availability and time, as to allow the development of the subsequent molecules to be put together in the ideal way as to be able to support life on plants, and do it so in a self-sustainable way, according to NASA. The theory that claims plants came first cannot and would not be able to explain this mystery, for it would seem that life wouldn't appeared on Earth if all the needed coincidences hadn't happened at the same moment.

But if we put melanin in the place of chlorophyll in the line of events that lead to what we know as life, things become a lot more coherent, if only for this: first, melanin can be synthesized abiotically, without intervention of a living organism. There's no arguing about it. There are many instances on nature, in Earth as in outer space, and in the whole universe. Plus, there are several ways to achieve it, with a critical mass of melanin in enough quantity, quality, availability, placing and stability, as to have an adequate and enough flux of chemical energy captured from photons- also in adequate quantity, quality, availability, placing and time – to let the other chemical reactions, also very important, that came from this very first reaction. Let's not forget that the order of these events remains the same till today, because otherwise life doesn't express itself, or does it in a poor fashion. In other words, the very first reaction implies an inevitable energy expense, completely covered up by melanin, which took it (and remains taking it, to this day) from electromagnetic radiations, always present in the environment, which comes from the sun in our solar system and other sources of the stellar space. Until today, such expense continues and will continue being covered by this energetic model.

Melanin at the Orion Nebulae, 1600 light years away
and 30 light years across.

Dark nebulae composed by melanin from Orion.

Dusty Iris Nebula of Melanin 1300 light years away

This means that, melanin absorbs all kinds of energy (all wavelengths, all frequencies and I believe also gravity radiations) and, once absorbed, melanin uses it mainly to unfold the water molecule, something totally unknown until now, and for it, unthinkable. The explanation of this "unknowing" is because melanin is considered "intractable" in a laboratory, for there is no way to study it in detail, because it resists to almost all analytic methods known, and it was almost "miraculously" that we had the fortune to discover at first it's wonderful photolytic properties (dissociating water absorbing light) watching carefully its extraordinary biological effects inside the human eye and few months later the second great property of melanin: it can support the opposite reaction, this is the water reformation.

Let's imagine, for a brief moment, Earth before life. The only things needed to begin the self-sustainable chemical system were light, water and melanin (the whole universe, and also on Earth, obviously). This can happen without many mysteries because hydrogen is the most abundant element in the universe. Earth is not the exception, even if it's not free in our environment, it's always found combined, especially with oxygen. Unlike other places in the universe where hydrogen has been found free, here on Earth mainly exists as a part of another both mysterious and wonderful substance, water. So the required elements to form water aren't something of a fantasy, something unbelievable or hard to explain in the environmental conditions of our planet.

The second element is melanin, which forms with amazing ease, and is also very stable, even in water, not only for thousands, but for millions of years. Thus, this association could happen all the times and/or during the necessary time, for the self-sustainable chemical system evolved, hatched, or in other words were able to begin a Darwinian evolution. Then, if electromagnetic radiations are omnipresent and water can be found in almost all corners of Earth, then melanin became the perfect complement for the existence of life on it.

It's not by chance that the zones with more availability of these three elements, indispensable for life are places where there has been found the most ancient specimens, like the Rift Valley in Africa.

Lets look at the past once more. Our planet was bathed day and night by a cloud of electromagnetic radiations in more than enough numbers for melanin to absorb it the most and began it endless activity, transforming photonic energy into chemical energy. This means melanin possess the amazing power to dissociate or unfold and reform the water molecule without pauses, day and night, for even in night there is enough energy on the environment, especially in the shape of several wavelengths from diverse electromagnetic radiations, enough for melanin to absorb them and immediately – in such a short time that's hard to imagine ($3X10^{-12}$ seconds) – begins to dissociate the water molecule.

Another property of melanin also appears: its extraordinary and incomprehensible (till today) capacity to moderate any energy it absorbs and to deliver chemical energy in a more stable and constant form, without the valleys and pikes that would have made such energy incompatible with life, for in some aspects, life only appears within narrow limits, let's say temperature, amount of water, amount of nutrients in the environment, etc..

This power to moderate sudden changes is of paramount importance to the origin and preservation of life, for it needs to be remembered that the sun's energy changes by the minute. This has been happening since the dawn of time, happens today and will continue to happen till the end of days and/or our solar system, for constant change is nature's imperative.

It we take notice from the phenomena occurring nearby our galaxy, or at least where energy is emitted as electromagnetic radiations, that may get to affect us, because is well known that we are eventually bathed by gamma rays (for instance) originated in the stars, things that had happened, happen usually and will keep happening.

We can talk about well known phenomena as supernovas, galaxies that "eat" another galaxies and things like that, that means, notable peaks in the surrounding energy, that, if not for the extraordinary properties of melanin, which absorbs it, uses, moderates it, avoiding peaks and lows beyond our tolerance, it's easy to understand that life on Earth wouldn't be as successful and diverse as we know.

What happens with this energy absorbed by melanin?

This relatively stable hydrogen carried energy supply extracted from water by melanin, put before nature an important amount of energy in lagoons, perhaps lakes, rivers or even larger water bodies, but with an omnipresent energy both day and night, given the fact that melanin is efficient enough to absorb electromagnetic radiations even at night, collecting enough energy even at night to keep a large enough energy supply to be considered meaningful. Besides, melanin is unlike chlorophyll, which only allows the unfolding of water, but is not capable to the opposite reaction, that is the rearming of water and with of electricity, as we show in the following chemical equations:

$$2H_2O \rightarrow 2H_2 + O_2$$

The above reaction is related to chlorophyll, but the reaction linked to melanin is as follows:

$$2H_2O \leftrightarrow 2H_2 + O_2 + 4e^-$$

Where is the key difference?

In chlorophyll the reaction's direction goes only to the right, but with melanin goes both ways. Chlorophyll only unfolds the water molecule, but it doesn't support its rearming. Melanin unfolds water (way more efficiently), but also supports the opposite reaction: oxygen reduction, this means melanin oxidizes water and also reduces oxygen. Chlorophyll only oxidizes water.

Why is this so transcendental?

Because if melanin only breaks down the water molecule then the oxygen levels would go up and up, eventually damaging any formed compound, whatever its name could be, it would even end changing melanin itself.

I don't think there could any arguing about this: too high oxygen levels are incompatible with life. However, melanin allows the reunification of hydrogen and oxygen, then the reaction gets balanced and the partial concentrations of the latter, at least inside the human body, never go above 97 or 98%.

Yet another argument that proofs the vital importance of oxygen levels not getting so high is chlorophyll, for nature put in it another kind of "locks".

The first one, is that a molecule who only has a 400 nm absorption range, wanes afterward and then shows a pike of maximum absorption around 700 nm, this means its absorption range is thousands times more restrained, more inefficient that melanin, meaning the generation of hydrogen and oxygen is also thousand times lesser.

Another "failsafe", is that plants can send oxygen molecules back to the atmosphere. They do this in such numbers that plants are considered the main oxygen source for the atmosphere. These two mechanisms seem to be enough to reassure oxygen's partial concentrations never reach high or dangerous levels to the same vegetable cell. However, evolution of these protection mechanisms surely didn't happen in the form one day to another; for they are way more elaborated than the ones melanin has to impede oxygen's highly destructive concentrations, already existing inside it.

Let's see it this way: melanin, since the first day the conditions for life to evolve – light, melanin and water - was able to protect it and the compounds slowly been added to it. It's a very less complicated theory about life's origin, with relatively less mysteries, a lot less complex, that if we try to explain the hatching of the cosmic imperative that is life beginning with plants.

An argument to be made about melanin being first, and chlorophyll second is the relatively frequent presence of melanin inside vegetables.

In the next image, we have an example:

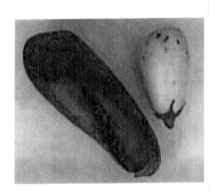 

Here we have two eggplant samples: the one from the left contains large amounts of melanin, comparing with the one on the right, which let us think two things: melanin is an advantage for plants as well as for animals, the size the eggplant with more melanin is an undisputed proof. The amount of melanin lets it to absorb a very superior quantity of luminous energy that the albino (melanin less) eggplant (right), that's why the difference in size is formidable. The second observation is a clear example of the existence of melanin in the plants world.

chlorophyll a R$_1$ = CH$_3$
chlorophyll b R$_1$ = CHO

In above figure, we have a and b chlorophyll's formula, very well known thanks to the actual laboratory methods that allow us to go deep in the molecule's structures. On the other hand, in the second above picture, we have melanin's (theoretical) formula. This let us to compare they share some very similar characteristics, for instance, the reaction center, formed by 4 hydrogen atoms, called 4N, which in chlorophyll have a prosthetic group: magnesium. However, melanin doesn't have prosthetic groups, reaction centers are free, and perhaps it represents an advantage in the experimental field.

But let's go back for a moment to the paragraph above: Melanin's formula is purely theoretical? How can this be? Doesn't have everything about it already been told? The answer is no. melanin is a molecule that, until today and despite the technological advances, which have been formidable, still keeps deep secrets to herself.

The former imagine does generate a doubt: melanin's presence in plants is relatively frequent, but examples of chlorophyll in the animal kingdom, are rare.

According to the latter, is simpler to assert, following the physical and chemical characteristics of these two molecules that melanin came before chlorophyll. I believe chlorophyll could have been derived from melanin, meaning organisms with melanin, eventually developed the skill to synthesize chlorophyll, but this organisms we call plants, grew a lot, that is, this kind of living beings, vegetables, had a lot of success, added to that, they're easier to observe.

There is another interesting data: one of the compounds closely related to the control of human photosynthesis, a tridecapeptide whose description would be included in this book's next edition, is one of the oldest compounds ever been proved scientifically and whose appearing goes back, even before than the immune system, and its existence has been scientifically proved in ghnastostomes and lampreys, that means, 700 millions of years ago. Consequently, this supports the hypothesis that melanin and not chlorophyll is the real precursor of life.

Then, the union of this three, already wonderful things, made the hatching of life on Earth possible: Light, Melanin and water – a unique liquid, because it's physicochemical properties, light, a seemingly never fully understandable phenomenon. Only the gathering of this three wonders could have generated another one, also a wonder itself: Life.

Without forgetting the single central substrate: The Universe, which we don't actually understand also, for it's not known where it begins, or where does it, ends. Something similar happens with melanin.

We must add here that oxygen concentrations inside melanin are unknown, but, in my experience, I've seen melanin giving out oxygen concentrations in about 34% of its surroundings, pretty safe concentration, far away from the deadly 100%. Besides, because of its physicochemical characteristics, melanin can regulate such concentrations and even get near the 100%, but never reaching it. A notable example of this, from my point of view, is the maximum oxygen concentration compatible with life, which in carefully controlled situations, especially inside the eye, is 97%. Hydrogen, as occurs in the whole Earth, is not alone inside the human body, it always can be found combined, for instance as water, which conforms 70% of the human organism generally.

Melanin also fits the theoretical requisites of metabolism being first. Melanin can be stable in water during millennia, breaking water over and over, incessantly, both at day as at night, is a fittingly stable energy source, not too much powerful, only enough to impulse in time and form, the initial reactions that brought life to this world. This initial reactions aren't too complex, but must be periodical, to allow the construction of the subsequent chemical reactions, which is also very important. A second requisite is that the released energy must impulse over and over several chemical reactions, but keeping them from being destroyed, the required time of it to keep them occurring over and over (the necessary number of occasions to made the path for the expression of life possible) inside a net of chemical reactions, who, in due time, will slowly step up its complexity, to allow adaptation and evolution (Darwinian evolution, of course).

The breaking and reforming of the water molecule is an oxidize-reduction reaction, that is, water get oxidized (loses hydrogen) then oxygen gets reduced (wins hydrogen). This is necessary for it to be considered the primary energy source par excellence. On the opposite hand, chlorophyll only oxidizes water, but it not allows its reduction afterward. Then, it cannot be an acceptable universal precursor of life, for its hydrogen supply with its valuable energy charge is rather restricted, barely enough for the plant itself, and if oxygen gets too high, it gets denaturalized, even chlorophyll itself can suffer from this. Besides, an energy supply that gets completely canceled at night, as occurs with chlorophyll, that doesn't get a single drop of energy by night, and thus, cannot unfold water, is less compatible with the generation of life.

Instead, melanin supplies energy day and night, because it is so efficient that even at night can collect enough energy to unfold water, certainly at a minor rate. This concept is more congruent with the generation of life, and perhaps this is the explanation of the need every living being has to sleep, because when melanin energy diminishes, activities also has to come to a minimum.

We may understand this better if we compare photosynthesis' energy with a digital camera battery. If the battery is low, all the camera's functions behave erroneously. Something similar happens inside the human body. If our battery is low (in this case, the energy we get from photosynthesis) then the body functions (all of them) behave erroneously, and depending on the subject former conditions, the period in which there is a diminished photosynthesis (because of cold weather, for instance), among many other variables, gives disease as a result.

And because everybody gets sick in his or her own way, then some people gets backache (back pain, low back pain, depending on the localization), some other gets Alzheimer's, some Parkinson´s, some multiple sclerosis, some arthritis, and so on.

Returning to the cell, we have amphipathic lipids, (which are soluble both in water and in lipids) that constitute the base of the cell membrane, a lipidic bilayer. These membranes circle the cells delimiting separations between the cell compartments. These bi-layers are mainly formed by phospholipids, and many of the cell's characteristics, what we call cellular biology, are founded upon this initial explanation.

Going back to water, in human physiology books we can find the following data: we drink water when we are thirsty, we also eat food to get energy from it, but food always has large amounts of water. Then, water, since it's a solvent, is used (as a vehicle) to excrete the waste products from body functions, to hydrate ions and to control body temperature trough sweat. It helps us to keep the blood volume stable, and so it plays a part on blood pressure. It is also the main ingredient of interstitial liquid and intracellular liquid.

But, being honest, the above data doesn't answer the question: why is water so indispensable to life? According to the above paragraph, water inside the body is only good to carry substances, in the blood stream, in urine, in sweat, inside the intestines, in cerebrospinal fluid, to name a few. It's hard to accept that this liquid, that only seems to support, to carry, to dissolve, as inner cleaner at most, is so indispensable to life.

And yet, however, is generally accepted that without water, there's no life, that life cannot express itself without water, and this goes for plants as for animals, how is it possible that without water, a human being dies so quickly, in a few days (about three days), depending on some variables like former physical condition, and environmental conditions, such as light and weather.

This doubt comes from the knowledge that, water, being the most powerful solvent known, is not the only one, and life could have been based on other solvents, as methane or ammonia. Perhaps on Earth this couldn't happen, because they exists as gases at room temperature and not as liquids, but inside other planets with the adequate conditions of pressure and temperature, such phenomenon could be possible.

H_2O

Chemical Formula: H_2O
Exact Mass: 18.01
Molecular Weight: 18.02
m/z: 18.01 (100.0%)
Elemental Analysis: H, 11.19; O, 88.81

CH_4

Chemical Formula: CH_4
Exact Mass: 16.03
Molecular Weight: 16.04
m/z: 16.03 (100.0%), 17.03 (1.1%)
Elemental Analysis: C, 74.87; H, 25.13

NH_3

Chemical Formula: H_3N
Exact Mass: 17.03
Molecular Weight: 17.03
m/z: 17.03 (100.0%)
Elemental Analysis: H, 17.76; N, 82.24

But this doesn't happen. Life as we know it doesn't express itself without water, it doesn't hatch if there's no water around. This works for both the vegetable and the animal kingdom. In plants, water's functions are similar: water as a solvent, as diluents, as volume, heat control trough evaporation. Until here, its functions are similar to the ones inside human body.

However, water's main function on plants, the generating of energy from sunlight, what we call photosynthesis, apparently didn't existed in humans and animals. Then, water generating energy through sunlight was only accepted (until now) in plants, cianobacteria and algae, but not in humans and/or mammals, there didn't seem to exist any example within animal kingdom, of this process, although it would be more correct to say we didn't saw it, because photosynthesis through melanin is very subtle process.

Thus, in plants the importance of the water molecule dissociation is very clear. However, if we consider that the concept of photosynthesis has been studied since approximately 350 years ago, and its understanding has been slowly discerned and clearly understood yet, we can appreciate it wasn't easy to perceive either.

Now, the description of photosynthesis is inside many books and is accepted without discussions that water is indispensable for life on plants, because it's the first (obligated) step in the series of chemical reactions which make life possible, what we call photosynthesis. This means, to obtain energy from water, there are at least three necessary elements – water, light and chlorophyll. Instead, inside animal kingdom there aren't so many examples of animals containing or expressing chlorophyll. Perhaps green caterpillars are an exception, but this only on insects.

The seeds of Papaw from Papaw tree, with a high content of melanin. The intrinsic property of melanin to split and re-form the water molecule explains the seeds germination, because when the water content of the seeds increases, then there is enough substrate to have an adequate output of energy, sufficient to impel the consequent biochemical reactions that lead to the blossom of the whole plant.

CHAPTER 10

Water as a solvent

Water is a strange and eccentric liquid, beside to be ubiquitous and commonplace; water is colorless, transparent, and tasteless. The ice floats because of the hydrogen bonding imposing a perfect tetrahedrally coordinated network[2], linking them into six-membered rings with much empty space between the molecules. This is the best known of what are widely seen as a long list of curiosities.

Water is H2O: two hydrogen's, each one attached to a central oxygen; the basic nuclear geometry of the molecule, with the bonded O–H distance of just less than 1 Å and H–O–H angle of about 104.5°. The water molecule is not a static entity; the molecule is in continuous internal motion, with the constituent atoms vibrating against each other.

Interactions between water molecules through the well-known hydrogen bond perturb symmetric and antisymmetric O–H stretch vibrations with respective frequencies of 453 meV and 466 meV, and a bending mode with a frequency of 198 meV from their isolated-molecule values.

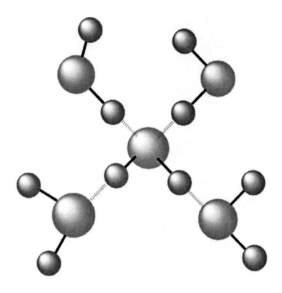

The molecule is made up from one heavy atom (oxygen) and two light atoms (the hydrogens). Therefore there is possibility that quantum effects might be relevant in some processes that involve water. Even at absolute zero, there is still zero-point motion.

The molecular chemistry depends on the electrons. And the water molecule has three vibrating nuclei; therefore we need to consider its electronic structure. Water has a high dielectric constant whose explanation is poorly understood.

A charge distribution is not only to do with the electrostatic attractions between molecules; it is also relevant to molecular repulsion. Water is a so small molecule.

There are always some H^+ and OH^- ions—or rather, their hydrated equivalents such as H_3O^+ or $H_5O_2^+$—in liquid water. Not only is molecular diffusion unexpectedly high in liquid water, but so also is the conduction of an excess proton. Water has a relatively thermally rigid and resilient network structure, yet one in which molecular motion is easily possible.

The strength of the hydrogen bond interaction between two water molecules has value of around 20 kJ mol–1 (ca. 5 kcal mol–1) that is about 10 times a typical thermal fluctuation at room temperature ($kT_{ambient} \approx 2.5$ kJ mol–1) Being significantly stronger than a typical van der Waals interaction.

Even the simplest molecular reorientation would be expected to require the breaking and reforming of at least two hydrogen bonds.

The maximum density of water is at 4°C and its unusually high thermal capacity is also familiar anomalies. Both the melting and boiling points of water are unexpectedly high when it is placed in the sequence of group VI hydrides.

If water is super-cooled very rapid the water fails to crystallize and becomes a glass[3]. This is very interesting because in its high density form it is the most abundant ice in the universe, where it is found as a frost on interstellar grains.

This is not the only regime in which water becomes amorphous. When cooled, water becomes exceptionally compressible, to extent that, if this property did not exist, the oceans would be about 40 m higher. The diffusivity of water is also anomalous, with the observation that initially as the pressure increases so doe's diffusivity.

Water is very biophilic, in spite that the intrinsic peculiarities of water are best understood by physicists. Biological functions are extremely sensitive to water properties. Even small doses of heavy water (D_2O) are known in at least some cases to be highly toxic.

The collective properties of water are central to both cellular mechanisms and macroscopic properties that range from its viscosity to planetary habitability. The organisms are largely composed of water under standard pressure and temperature[4]. However, even terrestrial life copes with water across a temperature range from well above boiling to below freezing, and ambient pressure from the rarefied heights near the top of the troposphere to the crushing environment of the oceanic trenches and within the Earth´s crust.

At elevated pressures and temperatures, water in some sense becomes less anomalous and less biophilic. The fine tuning of the specific interactions between water and biochemistry is evident. The interrelationships between water and the various organic substrates are so intimate that water itself must be treated as a biomolecule.

Water is an extraordinary solvent, as well as being far more than a solvent and cannot be considered in isolation. Water is the exact, perfect and unique substrate for melanin; otherwise its intrinsic property to split and reform the water molecule cannot be fully expressed.

Water also serves as a ligand. For instance, in both hemoglobin and cytochrome oxidase, the binding and subsequent release of water molecules is critical to their proper function. In the context of protein function the role of water is by no means completely understood.

The hydrophilic and hydrophobic interactions between water and a protein are crucial in successful folding, it is clear that the process is exceptionally finely balanced between competing demands. By one side proteins require stable folds, robust to environmental fluctuations, but by other hand must be flexible enough for allostery and complex interactions within the proteome while simultaneously exhibiting a rich and variable designability within the evolutionary random walk.

The free energy (ΔG) of the folded protein is remarkably low, being equivalent to two to four hydrogen bonds. The process of folding entails a very subtle range of weak, local interactions between molecules in an aqueous medium.

We still do not have a good understanding of water and its interactions with other molecules. The ability of water to exchange entropy with folding protein is one of its more astonishing properties. In spite to be strongly exothermic and therefore highly deleterious to cell function is, in terms of entropy, efficiently compensated.

The bonding strength between water molecules is enormously high, so explaining why it remains as a liquid at a much higher than expected temperature. This strength confers a sort of rigidity. There is little doubt that this combination of properties is important, perhaps critical, in protein function. Therefore the hydrogen bond in water is of fundamental importance.

Density and surface tension of water are properties that appear as relatively insensitive. Different properties exhibit varying degrees of sensitivity to changes in water geometry and hydrogen bond strength.

The hydrolytic activities of water present a severe barrier to the assembly of many molecules abiotically, not least the nucleotides. Alternative views insist that water is far from ideal, and that other fluids are strong candidates for alien life forms. However, water is the exact substrate for the intrinsic property of melanin to split and reform the water molecule. There is no other alternative therefore life would be literally unimaginable without water.

With the main aim to understand the strange dynamic entity we call life, the unraveling of the photo-system composed by Light / Melanin/ Water, arranged in order of abundance in the universe, allows the emergence of a general theory of organization.

The complexity, stability, robustness and sometimes extreme sensitivity of life are explained by the process that we appointed as human photosynthesis. The general prospect is that life is ubiquitous and is based in physicochemical systems derived all of them from the photo-system Light/ Melanin/ Water.

Beside of its close involvement with life, H_2O, mainly in its liquid state, is "strange and eccentric". The molecular properties of water that are of prime importance are: The sp^3 hybridization of the H2O molecular orbitals, which gives rise to an approximately tetrahedral disposition of four possible hydrogen bonds about each central oxygen atom. Water molecule has two positive and two negative charges, therefore its nature is quadrupolar. The low energies of the O-H-O (hydrogen) bonds, in comparison to covalent bonds or ionic interactions, make these bonds susceptible to minor structural distortions and surprisingly these are the only interactions in which water molecules can participate.

Although the above features are also found in other molecules, their combination in one substance is probably unique. The intimate relationship between water and life´s biochemical processes is demonstrated by substituting the deuteron (^{2}H) for the proton (^{1}H), substitution that merely a change in the zero-point energy. Even this minor isotopic substitution is toxic to most life forms. Therefore life processes are so sensitively attuned to the O-^{1}H-O hydrogen bond energies that the substitution by the heavier deuteron alters the kinetics of biochemical reactions. Only the lowest forms of life, in instance, some protozoa, can tolerate the complete deuteration of their constituent biopolymers, but only if brought about in gradual stages. All higher forms of life exhibit signs of enhanced senescence and eventual death.

Water remained an "element" until 1790, when Lavoisier and Priestly demonstrated that water could be decomposed into air (oxygen) and inflammable air (hydrogen). Much later the question of where water existed in the universe was established that most of it is adsorbed on interstellar dust particles that eventually make up the tails of comets.

The Earth water content, estimated as a total of 1.34×10^9 km^3, of which 97 % make up the oceans. The 99.997 % of the remaining fresh-water is locked in the Antarctic ice cap. The 0.003% represents the fresh water immediately accessible for agriculture, domestic, and industrial use. Of the total amount, 75 % is used for irrigation purposes.

The hydrological cycle: transpiration/ evaporation/, followed by precipitation; ensures that our surface water is recycled 37 times per year, constituting a vast water purification system. Unfortunately, 75 % of water precipitation falls into oceans and thus becomes largely useless, at least to the terrestrial animal and plant kingdoms, unless the salt is removed.

Within the overall hydrological cycle of transpiration and precipitation, two sub-cycles are between food producers (plants) and consumers (animals). The nitrogenous and phosphorus cycle are of almost equal importance; they can be easily upset where large quantities of fertilizers and/or detergents find their way into water sources, such as rivers and lakes. The imbalances caused between producers and consumers by excesses of nutrients, such as nitrates and/or phosphates, can have disastrous effects appointed as eutrophicacion (overfeeding) and the Lake Erie during the 1950s is sadly a very good example of this.

The interactions between water and molecules that govern life processes can be studied at several levels of increasing complexity. Over the past decades, much has been written about a so-called water structure. Biological and technological phenomena have been ascribed to this ill-defined water structure and to bound water in strongly or weakly way. Such distinctions run counter to the universally accepted laws of physics, which do not even recognize the existence of molecules at all.

Water can interact directly only by hydrogen bonding, either with ions or with molecules that, like water itself, possess proton donor and/or acceptor sites. Ionic hydration can, to some extent, be treated by the laws of electrostatics. Ionic hydration and hydration by direct hydrogen bonding can be treated by classical physical approaches.

Hydrophobic hydration describes the interaction between water and molecules or ions that "hate" water and are incapable of participating in the formation of hydrogen bonds. Such molecules include the noble gases and hydrocarbons; the simplest example is methanol in which the −OH group favors the interaction with water, but the $-CH_3$ group is hydrophobic and is repelled by water.

The balance is a result between hydrophobia and hydrogen bonding as to which effect will predominate. In the case of methanol, the −OH group wins, making the alcohol completely miscible with water. Molecules forming the cell membranes contain long alkyl chains and only one single polar head group. On balance they are therefore insoluble in water.

Folded, native protein structures are maintained by many stabilizing intrapeptide ionic and hydrogen bond interactions. These are, however, counterbalanced by destabilizing hydrophobic interactions between alkyl residues and water. The net stability margin of a protein in its active state rarely exceeds 50 kJ/mol; this supposes an energy equivalent to only three hydrogen bonds in a structure that contains many hydrogen bonds.

For a globular protein molecule to form a biologically active structure, it requires a ca. 50% content of (destabilizing) hydrophobic amino acids. It is thus the fine balance, caused by water-promoted interactions that have given us life's workhorses that are responsible for the majority of biochemical functions. However these concepts will change when beside the properties already known of water the energy released symmetrically in all directions are added to the equation.

We could say so far that biochemistry is the chemistry of water but from now in ahead will be the photochemistry of the Light/ Melanin/ Water, arranged in order of abundance in the universe. The versatility of the water is not enough to gives origin to the life. We must keep in mind that water is just a very important component of the universal photosynthesis system Light/ Melanin/ Water but necessarily the life's origin requires the complete photo-system; each separate component is unable to gives the very first spark of the life at all.

The H_2O molecule acts as proton-transfer medium in four basic types of biochemical reactions: oxidation, reduction, hydrolysis, and condensation. Melanin oxidizes the water and reduces the oxygen.

Below a threshold value, increasing pressure increases water diffusivity; for a normal liquid, the increased crowding results in the opposite behavior. The ensemble of bio-molecular processes requires an aqueous environment and energy.

The molecular weight (MW) of melanin is estimated in millions of Daltons, the MW of water is 18 Daltons; and the photon is massless. Energy can neither be created nor destroyed, and energy, in all of its forms, has mass. Mass also can neither be created nor destroyed, and in all of its forms, has energy. Whenever energy is added to a system, the system gains mass.

Into the cell everything is fine tuning processes. For instance, the native protein is only marginally stable, in order of 10-20 kcal mol $^{-1}$, which amounts to only two to four of the several hundred hydrogen bonds in a typical native protein solvent-system. Lose a small fraction of these hydrogen bonds and the native structure of the protein falls apart.

Each water molecule is capable of donating two hydrogen bonds through its two protons and accepting up to two through its negatively charged region. There is some evidence that a single accepted hydrogen bond is likely to be stronger than a fully four-coordinated molecule.

It is unlikely that the exposed protein surface would present equal numbers of hydrogen bond donors and acceptors to the surrounding solvent[5]; however one of the important properties of the water molecule is the ability to accommodate variable coordination. Water can accommodate itself to varying external hydrogen bonding requirements without its own structure being significantly affected.

Like liquid water, the protein is also held together in significant measure by hydrogen bonds[6]. By other side is interesting that in spite that the water network is relatively rigid allows a rapid molecular diffusion. The diffusivity should not be significantly compromised in the crowded conditions that are found in the cell.

The already known (and unknown) properties of water undoubtedly fit perfectly so the intrinsic ability of melanin to split and reform the water molecule can take place. The variability of the underlying ideal 2:2 donor/acceptor ratio of the water molecule seems to be crucial for melanin internal process and for life processes.

Many important biological processes including water dissociation and reforming; requires the transport of protons to or from a protein active site. Proton mobility in water is anomalously high[7]. Thus, water seems to be a particularly useful medium for facilitating this part of much biologically important process, photosynthesis included.

Melanin optimizes the water molecule properties but the deciphering of how and when will require a lot of work. The fine tuning inner characteristics of the universal photosynthesis system bring us to think that Light/ Melanin/ Water are the only way to generate life in the whole universe, and not only over the earth.

The interesting properties of water individually are demonstrated by other molecules; but only in water itself all these properties are found[8].

Water is the unique and universal matrix for life, it is a substance that actively engages and interacts with bio-molecules in complex, subtle, and essential ways. Therefore we could think that the active volume of molecules are beyond their formal boundary or the van der Waals surface because the shell of water that surrounds them is activated, that means that respond to the presence of the molecule.

The presence of an intruding solute particle modifies water's behavior and the nature of this response is far from obvious. Water is an extremely good solvent for ions, in part as a result of water's high dielectric constant. Water, also; is an efficient solvent for bio-molecular polyelectrolytes such as DNA and proteins. The electronic properties of water are the best fitted for the expression of the intrinsic capacity of melanin to oxidize water and reduces hydrogen.

Water molecules will solvate cations by orienting their oxygen molecules toward the ion, whereas they will adopt the opposite configuration for anions. Hydrophobic solutes in water experience a force that causes them to aggregate. It seems clear that this hydrophobic interaction is in some way responsible for several important biological processes[9].

However this single explanation is not enough to understand the aggregation of amphiphilic lipids into bilayer, the burial of hydrophobic residues in protein folding, and the aggregation of proteins subunits into multisubunit quaternary structures. It is possible that the alternate waves of diatomic hydrogen followed by reformed water accompanied by a relatively ordered flow of electrons that melanin releases in form of growing spheres, symmetrically in all directions; are the complimentary mechanisms of the so-called important biological processes.

I have no doubt that water alone has a very different molecular behavior compared with a system where melanin and water are combined. Even more; the electronic characteristics of both compounds change dramatically. It is ironic that water is one of the smallest molecules and melanin is the largest one.

The vital processes require a great number of ions (Latin: which travels) and molecules move in proximity, that is, to be soluble in a common medium. Water works as a universal solvent thanks to its two related properties: its disposition to form hydrogen bonds and its bipolar characteristic. Substances that can benefit from these properties are called hydrophilic or "akin to water". Let's not forget water is considered a highly polar liquid (it contains charges that make its combination with other charges possible, somehow like magnets: opposite charges are attracted to each other, similar charges are repulsed to each other.

Hydrophilic molecules inside a liquid solution means molecules with groups capable of forming hydrogen links or bridges usually will melt with water, on this kind of link. Thus, water easily dissolves hydroxyl compounds, acetones, sulfihydriles compounds, amines, esters; and a whole variety of organic compounds. It must be noted that, to combine them with water, they require charges, as with the former examples, charge less compounds, like lipids or fats, don't combine with water.

But not only givers and takers of hydrogen bonds and other kinds of links (covalent, Van der Waals forces, ionic, by electric attraction to name a few) are well dissolved by water.

Unlike most organic liquids, water is an excellent solvent for ionic compounds. The explanation for this lies in the bipolar nature of the water molecule. Interactions between water's dipoles with cations and anions inside an aquatic solution hydrate the ions, that is, they are surrounded by layers of molecules called hydration layers. This layer's formation is favorable energetically speaking (this means they are more likely to occur from an energetic point of view). Besides, water's high dielectric constant lowers the electrostatic force between opposite charged ions, that otherwise would be reunited.

The dipolar nature of the water molecule also favors water's ability to dissolve non ionic organic molecules although polar, like phenols, ester, and amides. Often, these molecules have high bipolar moments and the interaction with water's dipole favors its solubility in it.

Hydrophilic substances solubility depends on a favorable energetic interaction with water's molecules. So, is not strange that substances like non polar and no ionic hydrocarbons that cannot form hydrogen bonds, only show limited solubility in water.

Molecules which behave like this are called hydrophobic or "with aversion to water". However, energy is not the only factor that limits its solubility. When hydrophobic molecules are dissolved, they don't form hydration layers as hydrophilic molecules do. Instead, the water's regular net forms structures like clathrates (similar to a cage), who's better example is ice, or cages around non polar molecules. Water's molecules orderly disposition, which goes beyond the cage o clathrates shape, is produced by a lowering of the mix's entropy or randomness. The lowering of entropy favors low solubility of hydrophobic substances in water. It also explains the well known tendency of hydrophobic substances to conform aggregates on water.

We have all seen how oil makes droplets when we shake it inside vinegar. We need more order to surround two hydrophobic molecules with two separated cages than to surround them with one, hence, the hydrophobic molecules tend to group together.

There is a very interesting kind of molecules that show both hydrophilic and hydrophobic properties at the same time. Such amphipathic substances have a highly hydrophilic group as head, assembled to a hydrophobic tail, usually a hydrocarbon. When dissolving them on water, amphipathic substances form one or two layers in the water's surface where only group heads are under it.

However, when strongly shaken, micelles can be formed inside the solution (spherical structures formed by a single molecule layer) or bilayer vesicles. In these cases, hydrocarbon tails tend to arrange themselves in parallel shape, which allows them to interacting through Van der Waals forces. Group heads, ionic or polar, are strongly hydrated by the surrounding water.

Water, its strangest properties and the origin of Life

The fact that the intrinsic reactivity of water is undesirable for life is especially obvious when considering RNA and DNA. Cytosine, one of the four standard nucleobases in both DNA and RNA, hydrolytically deaminated to give uracil with a half-life of ca. 70 years in water at 300 K (Frick et al., 1987). Adenosine hydrolytically deaminated (to inosine), and guanosine hydrolytically deaminated to xanthosine at only slightly slower rates. As a consequence, water is constantly reacting with DNA in a way that causes it to lose its information; terrain DNA in water must therefore be continuously repaired.

Indeed, the reactivity of water, and the consequent destruction of genetic material, creates the biggest obstacle in understanding how life arose on Earth. Reaction of iron phosphates arriving via meteorites generates reactive phosphates that lead to the nucleotides (Pasek and Lauretta, 2005). Yet all is for naught if the RNA molecule is placed in water; it falls apart by hydrolysis.

In modern life, the aggressive reactivity of water with respect to molecules like RNA and DNA is managed using sophisticated repair systems. Such repair systems were presumably not present at the dawn of life. This creates a paradox in the structure of genetic matter at the dawn of life.

On one hand, the repeating backbone charge suggests that RNA must work in a hydrophilic solvent such as water. But the number of hydrolyzable bonds suggests that RNA was difficult to synthesize prebiotically in water.

This discussion summarizes only some of the issues that might be addressed as we attempt to combine the constraints of physical and chemical laws, the consequent behavior of solvents (like water), and the structure of biological molecules. Synthesis is an experimental approach to better understanding of this interaction, and especially for how underlying physical law has constrained biology at the molecular level. It also provides intellectual processes that manage the natural tendency for scientists to become advocates for their individual prejudices.

CHAPTER 11

The Visible and Invisible Light

It's a generally accepted fact that we could know more about the Universe if we understood light better than we do. Its name comes from Latin, lux, lucis and is usually described as group of electromagnetic waves, whose fundamental particle is called photon, and constitutes one of the main forms in which nature carries the energy from one place to another.

Normally, people understand light only as the visible electromagnetic radiations, that is, only those the human retina is able to capture. Photoreceptor cells inside it can only retain a small portion from the whole electromagnetic range, a thin part that goes from 380 nm (purple) until 780 nm (red) wavelengths (λ).

Visible light to the human eye is part of the spectrum colors and is ordered as the rainbow, conform the visible spectrum (above).

This is what we call visible spectrum; however, electromagnetic radiations go way beyond this, for radiation is defined as energy emission from a source, whether as waves of moving particles. Its properties will depend on the vibration's forms and frequencies, which are determined, among other things by the energy they carry.

Electromagnetic radiations are named, according to the above cited, as follows: gamma radiation, that goes from wavelengths whose size is lower than 10^{-9} m and its frequencies are 10^{-19} Hertz (Hz or times per second); immediately, we have X rays, whose wavelength is between 10^{-12} and 10^{-20} m and their frequency, that is, the times they go through a certain point on a given time, goes from 10^{17} Hz to 10^{20} Hz; then we have the ultraviolet radiations that goes from 10^{-9} to 10^{-7} wavelength, with a frequency that goes from 10^{15} to 10^{18} Hz. Now we have visible light with λ from 10^{-7} to 10^{-4} m with frequencies from 10^{14} to 10^{15} Hz. Next is infrared light, which, depending on circumstances, we can perceive as heat, with wavelengths from 10^{-6} to 10^{-4} m and frequencies from 10^{12} to 10^{14} Hz. The next electromagnetic radiation would be microwaves, with a λ that goes from 10^{-4} to 10^{-1}m and frequencies from 10^{9} to 10^{13}. At last will comes radio waves, whose λ is a meter long with a frequency of less than 10^{9} Hz.

Frequency and wavelength are related through this expression:

$$c = f\,\lambda$$

Where c is speed of light in vacuum, frequency f or v, and λ wavelength

---

Visible objects:

There are two kind of visible objects: those that give up light by themselves and those who reflect it. Their color depends on the light's spectrum and the object's absorption, which defines which waves will be reflected.

White light is produced when all the wavelengths from the visible spectrum are present in similar number and intensity. This can be verified in a spinning disc, with all the color equally distributed.

Human eye is sensible to this tiny range from the electromagnetic spectrum. Waves with lesser frequency than light (like radio frequency) have a bigger length and circle objects, not touching them. These allow us to have signal on our mobile phone even inside a house. Waves with a bigger frequency than light have a wavelength so small it goes through matter. X-rays for instance, goes through some materials, like flesh, although not through bones. Its only in the slice of the spectrum that goes from purple to red where electromagnetic waves interact with matter (reflection, absorption) and allow us to see objects, their shapes, their position, etc.. Inside this slice of the spectrum, we can say which frequencies are reflected, or emitted by any object, that means, what's its color.

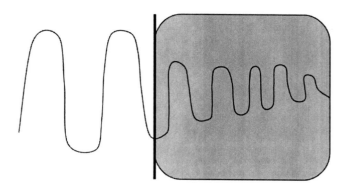

Wavelengths characteristics change when travel through different refraction index medium.

Wave Theory of Light

It is an old observation that two slits placed in a front of source of light give less light than just one slit. A phenomenon called interference. In order to see this phenomenon in which instead of being a small illuminated area we get a huge spread, a certain condition has to be satisfied as follows:

λ (Lambda) representing the wavelength, and it is much smaller than any distance between the borders of the slit. If the slit is thousand times the wavelength of light, then the shadow will be geometrical. But if we make the slit smaller and smaller until it becomes comparable to the wavelength, then you'll find it begin to spread out.

d is the distance between the slits, if it is much bigger than the wavelength and each slit is much bigger than the wavelength, then we will find that we cannot see interference effects. In fact the interference effects are always there, but they oscillate greatly on that screen, for instance 10 000 oscillations per centimeter, and the eye is able to look at it only one millimeter at a time, then the averaging is over a small region so we only find the average of the graph. Thereby, that looks like we got without interference.

In order to see interference, we need to make an experiment at a distance scale comparable to the wavelength. If we look at light using equipment with one or two holes or slits, but whose dimensions are much bigger than the wavelength, then we cannot see the wave theory of light, instead we will see the geometric theory.

But once we have an apparatus with a probe light at length comparable to wavelength, then interference phenomena will "happen". Typical wavelength light has 5,000 Angstroms or 500 nanometers (nm or $\mu\mu$) or 500 x 10^{-9} m, or around 10 million of oscillations of the wave within 1 meter. If we could have an instrument that can pick up that kind of fine features then we will able to see the wave nature.

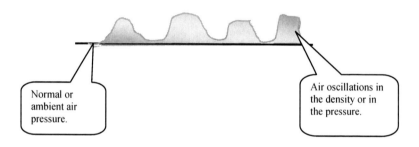

Normal or ambient air pressure.

Air oscillations in the density or in the pressure.

Presently only when people probed light with very tiny holes placed much closed each other, they saw it. Even if we made the holes very small, and if we shine white (polychromatic) light on the screen, we will not see any of these interference patterns. White light is made of many colors and many wavelengths, so each wavelength will form its own pattern. But the maximum and minimum of the eye will not coincide with that of the other. In order to see the interference effect we need monochromatic light sent by a source which is sending it a definite frequency and wavelength.

The nature of waves or light that produces these interference patterns

Light obeys a wave equation: $\Psi = \dfrac{d^2\,\Psi}{dx^2} - \dfrac{1}{V^2} = 0$

Ψ (Greek capital letter psi) generic name of whatever it is that is doing the oscillations. For instance, Ψ can be the height of some string that is vibrating.

Ψ can be E or B in the electromagnetic case or the reference level in a lake when the water does some oscillations. Ψ is the deviation from the normal. In the case of a sound wave or deviations in air pressure; for instance the graph would stand for the normal or ambient air pressure, and when we talk then the air has some oscillations in the density or in the pressure.

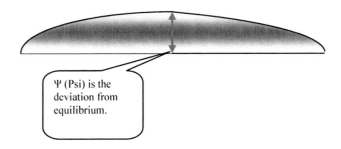

Ψ (Psi) is the deviation from equilibrium.

The context can differ but the equation is the same. The velocity will vary from problem to problem. In short notation the equation can be written down as follows:

$$\Box\,\Psi = 0$$

It is a linear equation which means if we multiply Ψ by a number it also satisfies the wave equation. Ψ can be positive or negative. However the brightness of life is always positive. We cannot have negative brightness. Thereby, in the electric field the brightness is not measured by the electric field, the one that oscillates, but by something quadratic in the field; and that is called intensity and is denoted by the capital letter I.

$$I \alpha \Psi^2$$

The intensity is proportional to square of whatever is oscillating. It is the measure of energy contained in the wave or brightness.

We can conclude this chapter asserting that the full spectrum of electromagnetic radiations goes way beyond what we can see, but, nevertheless, there are another radiations or energy forms that can only be detected with sensible devices for a specific wavelength. So the absence of evidence is not an evidence of absence.

CHAPTER 12

Melanin and Water, the Interaction between Them as a Result from Electromagnetic Irradiation

Water and melanin are molecules fundamentally neutral, because its pH is around 7. At the same time, they have a slight tendency to ionize, which means they can act as very weak acids or bases (they do not dissociate themselves much). One way to understand this ionization reaction (in water) is to watch a water molecule transfer a proton to another molecule to give an hydronium ion (H_3O^+) and hydroxyl ion (OH^-) as a result, in this way water is a taker and a giver of protons at the same time.

$$H_2O + H_2O \rightarrow H_3O^+ + OH^-$$

Actually, this is an oversimplification, because the moved proton can be united to several groups of water molecules such as $H_5O_2^+$ and $H_7O_3^+$. In a liquid dissolution, protons have plenty mobility and the charge jumps from a molecule to another in a split second, approximately 10^{-15} seconds.

To practical purposes, is always enough to describe the ionization process in a very easier way.

$$H_2O \rightarrow H^+ + OH^-$$

... never forgetting a proton is never free inside a liquid solution, never as free ion, but always associated with one or more water molecules. When we write a reaction with an H⁺ actually are talking about a hydrated proton.

The balance described in the former equation can be expressed in the shape of an ionic output from water, which is 10^{-12} M^2 at 25 °C.

$$[H^+][OH^-] = K_w = 1 \times 10^{-14}\ M^2$$

M means moles per liter. Since the ionic output is a constant, $[H][OH^-]$ it cannot vary independently from one to another. If we change $[H]$ or $[OH^-]$ adding acid or basic substances to the water, the other concentration must change accordingly. A solution with a high $[H]$, must have a low $[OH^-]$, and backwards.

If we use pure water, without any acid or basic substances in it, all H⁺ and OH⁻ ions must come from water's own dissociating. In these conditions, H⁺ and OH⁻ concentrations are the same, so at 25° C we have:

$$[H^+] = [OH^-] = 1 \times 10^{-7}\ M^2$$

So we say the solution is neutral, because is neither acid nor basic. However, the ionic output depends on weather, so a neutral solution does not always have $[H^+]$ and $[OH^-]$ at exactly 1×10^{-7} M^2. For instance, at human body temperature (37 °C) ionic concentrations $[H^+]$ and $[OH^-]$ inside a neutral solution are both $1.6 \times 10^{-7}\ M^2$.

The physiological values of the pH scale.

To avoid working on ten's negative potency, we almost always express hydrogen ion's concentration as pH, defined as:

$$pH = -\log [H^+]$$

The higher is $[H^+]$ inside a solution; the lower would be the pH, so a low pH corresponds to an acid dissolution. On the other hand, a low $[H^+]$ must go with a high $[OH^-]$, so a high pH, corresponds with a basic dissolution.

Most body liquids have values in a range from 6.5 to 8.0, in which is often called the physiological pH edge. Most biochemical processes do occur in this part of the scale.

Due to the biochemical processes' sensibility to even the smallest changes on pH, pH control is essential in most biochemical experiments. Most organic molecules answer to pH changes within the physiological edge or close to it; this often has a notable importance on their function.

Theoretically, melanin absorbs the whole electromagnetic spectrum, this is probably effectuated by melanin's peripheral parts, which is followed by a high energy electrons generation from low energy electrons. This high energy electrons go towards free radicals centers inside the compound, where they are probably taken by some element, like iron, copper, gadolinium, europium, etc.., from here they are transferred to a primary electron taker, whose nature is unknown until now, because it's a complex union, which comprehends ionic interactions. This electron transference releases energy, which is used to set a proton's gradient.

Combining water and melanin molecules conforms what we may call a photo system, which absorbs luminous energy, using it to, at least, two intertwined actions: to remove electrons from water and to generate a proton's gradient. Melanin's components are very close to each other, which makes easier a quick energy transfer. About 3 picoseconds after being illuminated, melanin's reaction centers responds moving a photo-excited electron towards the primary electrons taker.

This electron's transfer creates a positively charged giver and negatively charged taker. The importance of the two opposite charged species' creation makes itself apparent when we realize this two species' oxidize-reduction abilities, since one of them is electron deficient and can take in electrons, which makes him an oxidizing agent. On the other hand, the other compound has an extra electron, which can be loosed easily, making him a reduction agent. This event- formation of an oxidizing agent and a reduction agent from light- takes less than a second's trillion part and it's the first essential step to photolysis.

Due to being oppositely charged, this compounds show an obvious attraction to each other. Charge separation is stabilized (probably) by their own movement, at opposite sides inside the molecule. The negative compound is the first to give in its electron towards a quinone (Q1) and, possibly, this electron goes to a second kind of quinone (Q2), which produces a semi reduced quinone molecule, which can be strongly bonded to melanin's reaction center. With each transfer, the electron gets nearer towards melanin's reaction center.

Melanin's positively charged part is reduced, which prepares the reaction center to absorb another photon. The second photon's absorption sends a second electron along the way (negatively charged melanin towards the first and second quinone molecule – Q1 and Q2-). This second molecule absorbs two electrons, thus it combines with two protons. The protons used in this reaction could come from melanin molecule itself or from the surrounding water, causing the photo-system's hydrogen ions concentration to get low, which contributes to form a protons gradient. Theoretically, quinone reduced molecule dissociates itself in melanin's reaction center, being replaced by a new quinone molecule.

These reactions occurs at room temperature, but, when we modify the temperature, we can induce the reactions in one sense or another, depending on the other variables control – pH, magnetic fields, concentrations, gas partial pressure, electrodes, cells shape, etc..) - and depending too on the main purpose we want the process to have.

Water molecule separation in hydrogen and oxygen atoms is a highly endergonic reaction due to a very stable association between these two atoms. Water molecule separation (in hydrogen and oxygen atoms) inside a lab requires using a strong electric current, or rather to rise temperature up to almost 2,000 °C. This, (water electrolysis) can be done by melanin at room temperature, using only the energy from the sun, mainly between 200 and 900 nm wavelength, whether from a natural or artificial source, coherent or not, focused or disperse, mono or polychromatic.

The red-ox potential from quinone oxidized form is estimated approximately in +1.1 V, strong enough to attract the low w energy, firmly held electrons from the water molecule (red-ox potential +0.82), which's separates the molecule in hydrogen and oxygen atoms.

The water molecule separation by photo-pigments is called photolysis. The creation of a water molecule during photolysis, apparently, requires the simultaneous lose of four electrons from two water molecules, according to the reaction:

$$2H_2O \leftrightarrow 2H_2 + O_2 + 4e^-$$

A reaction center only can generate a positive charge or its oxidant equivalent at the time. The problem gets resolved, hypothetically, with the presence of 4 nitrogen atoms inside the melanin's reaction center, giving a single electron each.

This nitrogen concentration stores perhaps four positive charges transferring four electrons (one at the time) to the nearest quinone molecule. The nitrogen electrons transfer from the reaction centers towards the quinone gets done through a passage made with rest of positively charged tyrosine. After each electron gets transferred to a quinone+ regenerating quinone, pigment is deoxidized (again to quinone⁺) after absorbing another photon to the photo-system. Hence, storing four positive charges (oxidant equivalents) nitrogen atoms from the reaction center gets modified by successively absorbing four photons by the melanin photo-system. Once the four charges are stored, the oxygen releasing quinone complex can catalyze, removing 4e⁻ from $2H_2O$, forming an O_2 molecule and regenerating the completely reduced nitrogen store from the reaction center.

Photolysis' produced photons are released in the milieu, where they contribute to the proton's gradient. The photo-system must be illuminated several times before releasing O_2 and thus hydrogen able to be measured, this indicates the effect from individual photoreactions must be stored before hydrogen and O_2 gets released. Quinones are considered as carriers for mobile electrons.

Let's not forget all electrons transfers are ex-ergonics and happen as electrons are transferred to carriers with a growing affinity to them (more positive red-ox potentials). The necessity of mobile electrons carrier's presence makes itself apparent. Photolysis generated electrons can go to a number of inorganic takers, which gets reduced by this. This electrons paths can lead (according to the used mix composition) to an eventual reduction of nitrates molecules (NO_3^-) into ammonia molecules (NH_3^-) or sulfates into sulphydriles (SH^-), reactions that turn inorganic waste into compounds needed for life. Thus, sunlight energy can be used not only to reduce carbon atoms into its most oxidized shape (CO_2) but also to reduce nitrogen and sulfur to its most oxidized shape as well.

The production of an O_2 molecule requires four electrons to be removed from two water molecules. The remove of four electrons from water needs absorbing four photons, one for each electron.

Melanin's electrolyzing properties (among many other properties) can explain the pike generated by visible light on an electro-retinogram, because when melanin lights up, intracellular pH descends, which activates chlorine channels sensible to pH inside the lateral base membrane. (The light pike is a rising in potential that follows FOT (fast oscillating though) phase and forms the slower and longer component from the direct current electro-retinogram (Kris 1958, Kolder 1959, Kikadawa 1968, Steinberg 1982).

Melanin, melanin's precursors, melanin's derivatives, melanin's variations and analogues oxidize the water molecule into O, O_2 and H_2, consuming energy from light (photons) and reducing oxygen atoms with hydrogen atoms into H_2O molecules, releasing energy (releases electricity, although it can also "store" electricity, it too can work as a battery or accumulator. In other words, not only does it generates energy, but also is able to keep it for a long time, within some, not so well defined yet, limits.

Photosynthesis' effects upon the eukaryotic cell's oxidize-reduction capacity are very interesting. Experiments show that when we stimulate the photosynthetic effect in melanin, this means, and its extraordinary ability to unfold water and hence to rise inside the cell the highly valuable energy carrier hydrogen, the cells red-ox ability increases fourfold, depending on the time we keep this photosynthesis stimulation and how do we do it. This is why when we treat diseases like Alzheimer, arthritis, colitis and others; the improvement grows while the patient stays in treatment. This means the physiological and/or therapeutic effects doesn't seem to have a limit. Besides, when we make such stimulation upon the eukaryotic cell, it becomes more selective to capture nutrients from its surrounding environment. It becomes more careful, more demanding, hence, more diseases can be prevented when photosynthesis is stimulated by prophylactic means.

Another curios fact is related to sleep. Until now, nobody knows why all mammals have the necessity to sleep.

However, if we consider day and nights' energy difference on the environment, or the difference in available radiated energy, then it seems sensible to propose that, as life originated with this same rhythm, then, when there is a lesser amount of available energy (in other words, less hydrogen, and, thus, less oxidize-reduction capacity inside the cell) by night our abilities descend notably, being at its height by day, when the biggest amount of available environmental energy is out. At night our skills descends notable, since the cell doesn't has all its oxidize-reduction (red-ox) ability.

The cell's red-ox ability is a basic element for life; it's the equation for life. First was melanin, thanks to its ability to transform visible and invisible light's energy into chemical energy, then there was an initial red-ox chemical reaction, given the fact water unfolding is an example of oxidize-reduction and the subsequent reactions, slowly concatenating with each other to express what we call life, are nothing but yet another oxidize-reduction examples. One more proof exists in plants, which also support their life form based on this kind of chemical reaction. For it to occur, flowing energy cycles are required, because, interestingly, when plants are seeded on controlled conditions, as to be constantly irradiated, the plant obviously grows faster, but with a very unpleasant taste. This is the reason why it's not a widely used practice.

We can describe the oxidize-reduction reactions as a matter exchange between two substances, molecules, cells, stuff like that, that would always occur simultaneously. This means, for a molecule to lose a part of it, electrons for instance, another substance has to win it. One cannot happen without the other. This matter exchange can be at the electron level (one or many), or at complete atoms, and so forth, since it depends on several factors as context, the amounts of reacting substances, light, etc..

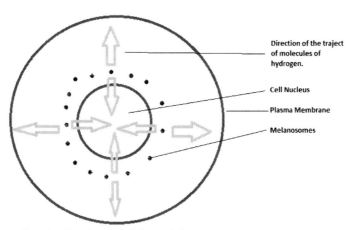

Direction of the traject of molecules of hydrogen.

Cell Nucleus

Plasma Membrane

Melanosomes

The center of the cell nucleus is a high concentration zone of diatomic hydrogen due to the confluency of the molecules of hydrogen that are released symmetrically in all directions by the melanin of the melanosomes and therefore is a high energy area.

Melanin is the human equivalent to plants´ lignin, and hemoglobin is the human chlorophyll.

Melanin and lignin are able to split and re-form the water molecule, hemoglobin and chlorophyll is able to split irreversibly the water molecule.

CHAPTER 13

Why melanin works day and night?

White lights are produced when all wavelengths from visible spectrum are present, in equal proportions and intensities. This can be verified on a fast spinning disc with all colors uniformly distributed on it.

White light is what we call light, because human eye is sensible to this small range from the radio-electric spectrum. Waves with lesser frequency than light (like radio frequencies) have a bigger wavelength and goes around objects not interacting with them, as happens with mobile phones.

Waves with bigger frequency than light have such a tiny wavelength that they go through matter, as do x-rays, which go through flesh, but not bones. Only in spectrum's part that goes from purple to red, where electromagnetic waves interact (being reflected or absorbed) with matter and allow us to see objects, its shapes, its position, etc... Inside this part of the spectrum we can find out the frequency or frequencies each object reflects or absorbs, which means, knowing its color.

Then, during night most people assume there is no light, which is truth but only to a certain point, because there is no white light, but another wavelengths are present, even if we don't see them. Our senses deceive us on this. There are wavelengths that don't interact with objects, which mean they don't go absorb nor reflect them, and this includes Earth itself. This means, there are wavelengths that goes through the Earth, without losing any of its contained energy.

And we're not even mentioning another energy forms we barely glimpse, like gravity or another, unknown energies. Let's just say the Big Bang theory, after a few years reign, is on retreat, thanks to the finding of the universe's expansion, giving its place to the fundamental chords theory, even if it's far away of being completely understood.

We are being repeatedly bathed by radiations, both by day and by night and being melanin as efficient as it is, is not that hard to concede it can produce energy even by night.

In other words, melanin captures enough energy, both by day and night, to generate useful flow of electrons, which, when putting it to a practical use, and when massive lighting for buildings and houses with melanin can become a reality, would be possible without using the dangerous lead-acid accumulators, something still not possible on solar panels, because these designs don't produce any energy at all by night, that's why by day, when they have an efficient performance, energy must be stored, and apparently there's no other known way to do so. We have to remember that until today, centralized energy generation has an unsolvable problem: produced energy either is immediately wasted or lost forever.

Those first molecules and/or structures that were the origin of life can resist partial oxygen pressures that are not tolerated otherwise on these days.

However, another tricky question rises due to this observation: plant cells break the water molecule and send oxygen back to the atmosphere. Thus, chlorophyll molecule, something so delicate, so extraordinary difficult to achieve as a substance with the extraordinary capacity to unfolding water molecule was created only to gives oxygen back into the atmosphere? No, it wasn't, because for this to be so, we would have to concede that nature does useless, trivial or confuses things, which is hard to accept. A more sensible answer would be to assert the breaking of the water molecule as the first self-sustainable biochemical reaction, for it feeds from radiations coming from the sun itself, getting the highly valuable hydrogen from it. This makes everything more congruent, for it's really hard to say: before, things were this way, now is completely different.

The theory that says plants came first sounds hard to believe, likewise with the argument about the transition between plant lives to animal life. I figure it was something truly fantastic when the first three walked, and yet again, an enigma, why this wondrous event hasn't repeated itself until these days? It's unbelievable to imagine this happening only enough times (which had to be thousand times in the required duration) along millions of years and never to occur again.

What it is that makes hydrogen so valuable? The energy it carries the same remains coherent millions of years since. Hydrogen is not nature's energy carrier by default, even when it only carries it, generating nor has energy at its own. This means, hydrogen needs the energy to be previously injected on it.

Consequently, thanks to hydrogen, and somewhat less thanks to oxygen, life, both vegetable and animal exists. The breakthrough represented by understanding melanin also captures photonic energy, and ultimately, breaking the water molecule it transforms it into chemical energy – which can be used by living beings, or when life was beginning by unicellular microorganism- allow us to conclude melanin, because its physicochemical characteristics is a more congruent precursor of life, one that keeps until today the same characteristics which made possible its original role. To think melanin could have brought origin to life seems more sensible than think chlorophyll was. Let's put forward a powerful reason: melanin exists even before life itself, is a highly stable compound (million years) and can be formed spontaneously (something that chlorophyll can't).

How do I support the former asserts? Melanin has been found in outer space in the shape of particles called cosmids. This finding allows to understand, to support, that melanin forms spontaneously even in the extreme conditions of outer space, with such extreme temperatures, it endless movement, unprotected irradiation, and despite this melanin shows an extraordinary stability, it can last eons.

Yet another data: melanin usually goes upon ice or froze water. This means, two compounds that, accordingly to my theory, gave origin to life, they are together very often and they both can withstand extreme conditions.

Both melanin and water can be found in outer space, withstanding total, or nearly total vacuum, and temperatures that go from -152°C to something above 120°C, plus the rain from powerful electromagnetic radiations from different wavelengths. Both molecules stability is pretty much out of discussion.

The fact they are wandering through the cosmos is compatible with the idea that when they find the adequate conditions, similar to our planet, life will hatch, upon this, we can accept the phrase that claims life is a cosmic imperative.

The biological process replicated in the laboratory.
These LEDs luminaries are energized with melanin.

Chapter 14

The Color of the Skin

Reasons for the different colors of skin have been tried to be explained since ancient times. Racial issues have been advocated, besides the advance in the human being evolution, or even the degree of civilization´s development, but more important, the most consistent finding is the correlation between the amounts of sunlight and skin´s color in the geographical where the people live.

In coldest countries, where the sunlight levels are lowest, the people trend to have fair skin, but the main reason is to keep the amount of light visible and invisible that reaches the inner body within certain range or limits, so the amount of photonic energy available inside is enough to maintain adequately all the cellular functions.

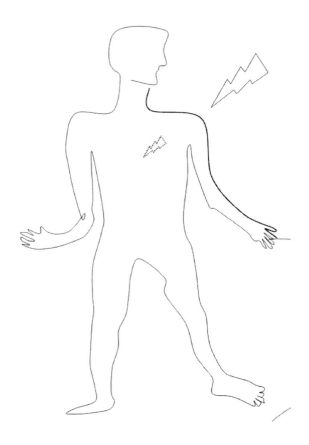

By other side, in places with intense sunlight levels, the skin color is noticeable darkened, in order to reduce the amount of photonic energy that reaches the inner parts of the body, avoiding excessive levels of photonic energy, and therefore keeping healthy the limits of radiation.

Therefore the main function of skin's color, is in regards energy, and is the modulation of the photonic energy goes through it.

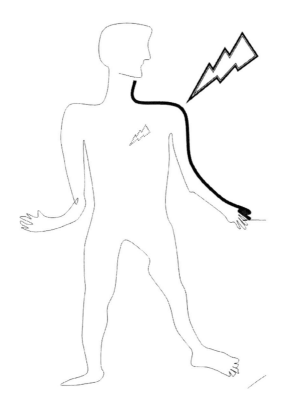

The changes in color of skin, that could be transient or more permanent, this regulatory mechanism is effective within certain limits, so the human body has other adaptive cellular behavior, for instance in fair skin the mitochondria´s number increase even until 83 % more than in comparison with dark skin subjects.

This is the explanation why the skin´s color is a character highly conserved in evolution. It is very important that the amount of photonic energy available in the inner side of the body bee in a good level and keep still. These are extensive to mammals, for instance, in mouse, the furrow´s color is determinate by more than 100 genes.

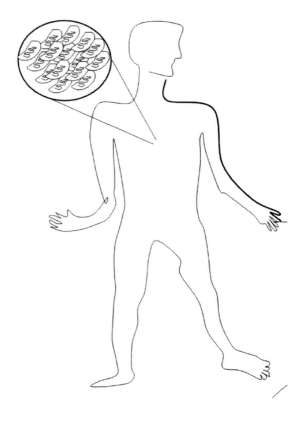

With fair skin the number of mitochondria increase until 83%.

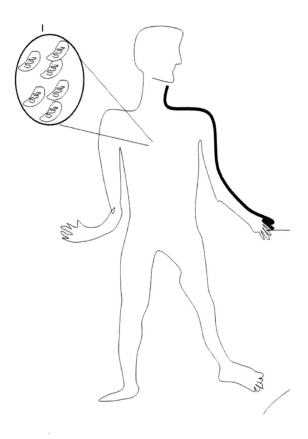

With darkest skin the mitochondria´s number diminish significantly.

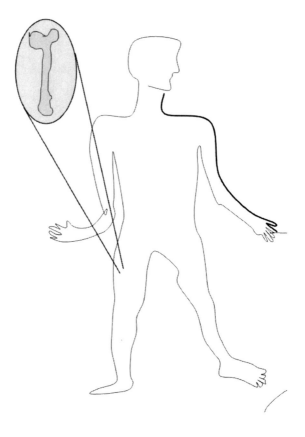

Fair skin means thinnest bones.

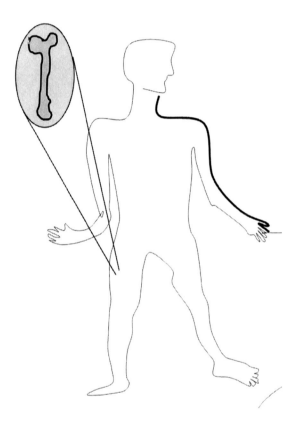

Dark skin means stronger bones.

African Americans, 25-hydroxyvitamin D, and osteoporosis: a paradox[1-4]

John F Aloia

ABSTRACT

African Americans have lower serum 25-hydroxyvitamin D concentrations and a lower risk of fragility fractures than do other populations. I review the evidence on factors other than vitamin review the calcium economy in each life seg Americans.

The distribution of serum 25(OH)D concent African Americans and whites from the third Na

different life stages. Researchers are actively trying to explain this genetically programmed advantage. Factors that could protect African Americans against fracture include their higher peak bone mass, increased obesity rates, greater muscle mass, lower bone turnover rates, and advantageous femur geometry. In addition, bone histomorphometry in young adults shows longer periods of bone formation. Although African Americans fall as frequently as do whites, the direction of their falls and their manner

Am J Clin Nutr 2008;88(suppl):545S–50S. Printed in USA. © 2008 American Society for Nutrition

Chapter 15

CO₂ and Human Photosynthesis Process

How CO₂ is produced in the cell?

Firstly the CO_2 is the most oxidized form of carbon. C=O=C is the final product of the oxidation of glucose and fatty acids in presence of oxygen. We could say that the formation of CO_2 is result of the complete aerobic oxidation (burning) of glucose and is coupled to the synthesis of as many as 36 molecules of ATP.

$$C_6H_{12}O_6 + 6\ O_2 + 36\ P_i^{2-} + 36\ ADP^{3-} + 36\ H^+ \longrightarrow$$
$$6\ CO_2 + 36\ ATP^{4-} + 42\ H_2O$$

Most eukaryotes are obligate aerobes: they grow only in the presence of oxygen –and light—and metabolize glucose (or related sugars) completely to CO_2, with the concomitant production of a large amount of ATP, compound that is toxic in large quantities and must be broken down to ADP in less of 20 seconds, otherwise ATP will disturb in a generalized form the highly complex metabolic processes developed along four billion years of evolution. When ATP is broken down to ADP requires energy, therefore absorbs more than release it.

Most eukaryotes can generate some ATP by anaerobic metabolism but the presence of light visible and invisible is an unavoidable and vital requirement. A few eukaryotes are facultative anaerobes: they grow in either the presence or the absence of oxygen; but certainly not in absence of light. For example, annelids, mollusks, and some yeasts can live and grow for days without oxygen but the ever presence of electromagnetic radiations is unavoidable and essential. The evolution was in that way.

Certain prokaryotes are obligate anaerobes: they cannot grow in the presence of oxygen, and they metabolize glucose only anaerobically, however visible and invisible light is a must.

In the absence of oxygen, glucose is not converted entirely to CO_2 as it is in obligate aerobes but to one or more two- or three-carbon compounds, and only in some cases to CO_2. For instance, yeasts degrade glucose to two pyruvate molecules via glycolysis, generating a net of two ATP and two NADH molecules per glucose molecule, but ATP is a toxic compound that must be degraded to ADP in less of 20 seconds, otherwise increasing amounts of ATP will impair gradually also the fine tuned intracellular processes developed by Nature along four billions years of evolution in presence of a full electromagnetic spectrum during night and day.

By other side, reduction of NAD+ to NADH is just one of the many ways that the cell uses the diatomic hydrogen released by melanin. Diatomic hydrogen didn´t combine with water.

Pyruvate dehydrogenase, a very large, multienzime complex in the mitochondrial matrix converts pyruvate into acetyl CoA and CO_2.

The final stage in the oxidation of glucose entails a set of nine reactions in which the acetyl group of acetyl CoA is oxidized to

CO_2. These reactions operate in a cycle that is referred to by several names: the citric acid cycle, the tricarboxylic acid cycle, and the Krebs cycle.

Acetyl CoA plays a central role in the oxidation of fatty acids and many amino acids. In addition, it is an intermediate in numerous biosynthetic reactions, such as the transfer of an acetyl group to lysine residues in histone proteins and to the N-termini of many mammalian proteins. Acetyl CoA also is a biosynthetic precursor of cholesterol and other steroids. In respiring mitochondria, however, the acetyl group of acetyl CoA is almost always oxidized to CO_2.

Mitochondria are among the larger organelles in the cell, each one being about the size of an *E. coli* bacterium. Most eukaryotic cells contain many mitochondria, which collectively can occupy as much as 25 percent of volume of the cytoplasm. They are large enough to be seen under a light microscope, but the details of their structure can be viewed only with the electron microscope[10].

Mitochondrial oxidation of fatty acids is the major source of ATP in mammalian liver cells, and biochemists at one time believed this was true in all cell types. Experiments in rats suggested that peroxisomes, as well as mitochondria, can oxidize fatty acids. These small organelles, $\approx 0.2 - 1$ μm in diameter, are lined by a single membrane. Also called *microbodies*, peroxisomes are present in all mammalian cells except erythrocytes and are also found in plant cells, yeasts, and probably most eukaryotic cells. The peroxisomes are now recognized as the principal organelle in which fatty acids are oxidized in most cell types. Indeed, very long chain fatty acids containing more than about 20 CH_2 groups are degraded only in peroxisomes; in mammalian cells, mid-length fatty acids containing 10 – 20 CH_2 groups can be degraded in both peroxisomes and mitochondria.

Most enzymes and small molecules involved in the citric acid cycle are soluble in aqueous solution and are localized to the matrix of the mitochondrion. This includes the water-soluble molecules CoA, acetyl CoA, and succinyl CoA, as well as NAD^+ and NADH.

The protein concentration of the mitochondrial matrix is 500 mg/ml (a 50 percent protein solution), and the matrix must have a viscous, gel-like consistency. When mitochondria are disrupted by gentle ultrasonic vibration or osmotic lysis, the six non-membrane-bound enzymes in the citric acid cycle are released as a very large multiprotein complex. The reaction product of one enzyme, it is believed, passes directly to the next enzyme without diffusing through the solution. However, much work is needed to determine the structure of the enzyme complex: biochemists generally study the properties of enzymes in dilute aqueous solutions of less than 1 mg/ml, and weak interactions between enzymes are often difficult to detect.

CO_2 is the final stage of the respiratory cycle and it merely needs to be exchanged with oxygen, however the function of carbon or carbon dioxide is as a carrier or acceptor molecule for the diatomic hydrogen that is released by melanin. There is free gas in some biological reactions and NO is a good example beside the diatomic hydrogen and oxygen released by human photosynthesis.

Our blood contains CO_2, usually around 40 mmHg, which is close to 5 %. The majority of CO_2 exists in the blood in the form of bicarbonate (HCO_3^-) which acts as a pH buffer to allow for gas, nutrient and metabolites fluctuations without causing wild pH changes. The Henry's Law: The solubility of a gas in a liquid depends on temperature, the partial pressure of the gas over the liquid, the nature of the solvent and the nature of the gas. The most common solvent is water. Gas solubility is always limited by the equilibrium between the gas and a saturated solution of the gas. The dissolved gas will always follow Henry's law.

The concentration of dissolved gas depends on the partial pressure of the gas. The partial pressure controls the number of gas molecule collisions with the surface of the solution. If the partial pressure is doubled the number of collisions with the surface will double. The increased number of collisions produces more dissolved gas.

When a gas is in contact with the surface of a liquid, the amount of the gas which will go into solution is proportional to the partial pressure of that gas. A simple rationale for Henry's law is that if the partial pressure of a gas is twice as high, then on the average twice as many molecules will hit the liquid surface in a given time interval, and on the average twice as many will be captured and go into solution. For a gas mixture, Henry's law helps to predict the amount of each gas which will go into solution, but different gases have different solubilities and this also affects the rate. The constant of proportionality in Henry's law must take this into account. For example, in the gas exchange processes in respiration, the solubility of carbon dioxide is about 22 times that of oxygen when they are in contact with the plasma of the human body.

When gases are dissolved in liquids, the relative rate of diffusion of a given gas is proportional to its solubility in the liquid and inversely proportional to the square root of its molecular mass. Important in the transport of respiration gases is the relative diffusion rate of oxygen and carbon dioxide in the plasma of the human body. Carbon dioxide has 22 times the solubility, but is more massive (44 amu[11] compared to 32 for oxygen). According to Graham's law, the relative rate of diffusion is given by

$$\frac{diffusion\ rate\ of\ CO_2}{diffusion\ rate\ of\ O_2} = 22\sqrt{\frac{32}{44}} = 19$$

The net diffusion rate of a gas across a fluid membrane is proportional to the difference in partial pressure, proportional to the area of the membrane and inversely proportional to the thickness of the membrane. Combined with the diffusion rate determined from Graham's law, this law provides the means for calculating exchange rates of gases across membranes. The total membrane surface area in the lungs (alveoli) may be on the order of 100 square meters and have a thickness of less than a millionth of a meter, so it is a very effective gas exchange interface. (Fick's Law)

Hemoglobin, the main oxygen-carrying molecule in red blood cells, can carry both oxygen and carbon dioxide, although in quite different ways. The decreased binding to carbon dioxide in the blood due to increased oxygen levels is known as the Haldane Effect, and is important in the transport of carbon dioxide from the tissues to the lungs. Conversely, a rise in the partial pressure of CO_2 or a lower pH will cause offloading of oxygen from hemoglobin. This is known as the Bohr Effect.

If pure water were on both sides of the membrane, the osmotic pressure difference would be zero. But if normal human blood were on the right side of the membrane, the osmotic pressure would be about seven atmospheres! This illustrates how potent the influence of osmotic pressure is for membrane transport in living organisms. It is a rising concentration of **carbon dioxide** — not a declining concentration of oxygen — that plays the major role in regulating the ventilation of the lungs.

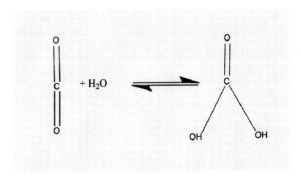

Carbonic anhydrase enzyme is able to catalyze one million of molecules of CO_2 by second.

Lung Physiology

Normal Lung physiology is unfortunately extremely complex, and this complexity is further enhanced in sick lungs. We tend to rely on gross over-simplification. The relationship between our current understanding of how lungs function, rely in many fruitful analogies that we use that gave been turned into dogma.

Background:

Atmospheric oxygen arose as a toxic by-product of the very first photosynthetic organisms, which were possibly quite similar to today´s blue-green algae. Organisms rapidly learned to use this oxygen, and minimize its toxic effects. When they began to co-operate to form multi-cellular organisms then they needed:

1. - Processes for removal of carbon dioxide, the principal metabolic waste product.

2. - Some ways to optimize deleterious levels of oxygen that comes from melanin water splitting and reform activity and the atmospheric levels. Hemoglobin was the answer. The red color means that hemoglobin absorbs the other wavelengths and the red´s wavelengths are not absorbed, therefore are reflected. Hemoglobin is the real equivalent to the chlorophyll and lignin is the equivalent to the melanin. Pyrrolidins are those compounds that make green the plants and red the blood.

$$2H_2O \rightarrow 2H_2 + O_2$$

This is the same reaction that happens in chlorophyll and hemoglobin.

We think that hemoglobin was the best answer to the oxygen presence at least by two things: By one side hemoglobin has the capacity to carry on oxygen at the adequate levels so the basic human photosynthesis reaction:

$$2H_2O \leftrightarrow 2H_2 + O_2 + 4e^-$$

Constantly tends to the left of the reaction in tissues. By other hand, the higher levels of CO and CO_2 drive the reaction to the right.

The physicochemical characteristics of hemoglobin and the tightly entwined cardiovascular and respiratory systems reflects clearly 4 billion years of evolution, therefore our abstraction capacity is easily overwhelmed when we are trying to understand the system patiently developed by Nature.

The respiratory system is an efficient pump for passing air over the capillary bed of the lung, where oxygen moves into the blood and CO_2 is removed from the blood, both by simple diffusion. The inefficient and failure-prone cardiovascular system is turned very efficient if we take in account that the energy produced by the intrinsic property of melanin to split and reform the water molecule generates several kind of forces insides the blood vessels that finally resemble a powerful vacuum that drive the blood stream along the blood vessels in a much more important way that the relatively weak contraction force of the left ventricle, it´s enough to think that 120 or even 150 mmHg are not able to push an erythrocyte through 95 000 km of capillaries in less of five minutes, indeed must be something else like this vacuum-generate forces. The heartbeat is only the drops that fill-up the glass.

The oxygen cascade is the name of several processes which documents the changes in partial pressure of oxygen from inspired air down to the mitochondrion where the oxygen is actually used. Oxygen moves down a gradient from a partial pressure of about 160 mmHg in the atmosphere, down to about 4-20 mmHg in mitochondrion.

Oxygen is taken up in the lung. This will decrease the amount of oxygen in the alveolar air in approx. 3 mmHg. The decrease will be directly related to the amount of oxygen taken up, and inversely related to the alveolar ventilation. In other words, the greater the alveolar ventilation, the less the effect of this oxygen uptake on the fraction of oxygen in the alveolar air beside the oxygen released by the water splitting property of melanin. Perhaps the main difference will be in the night, when the amount of light visible and invisible is naturally decreased and therefore the rate of the reaction will be slightly different between night and day.

The big drop comes at tissue level, where PO2 within mitochondrion has been estimated to be as low as 4-20 mmHg. In some normally functioning cells this PO2 may even drop to 1 mm Hg.

The lung architecture

Surfactant is a complex substance containing phospholipids and a number of apoproteins. This essential fluid is produced by the Type II alveolar cells, and lines the alveoli and smallest bronchioles. Surfactant reduces surface tension throughout the lung, thereby contributing to its general compliance. It is also important because it stabilizes the alveoli. Laplace's Law[1] tells us that the pressure within a spherical structure with surface tension, such as the alveolus, is inversely proportional to the radius of the sphere (P=4T/r for a sphere with two liquid-gas interfaces, like a soap bubble, and P=2T/r for a sphere with one liquid-gas interface, like an alveolus: P=pressure, T=surface tension, and r=radius). That is, at a constant surface tension, small alveoli will generate bigger pressures within them than will large alveoli. Smaller alveoli would therefore be expected to empty into larger alveoli as lung volume decreases. This does not occur, however, because surfactant differentially reduces surface tension, more at lower volumes and less at higher volumes, leading to alveolar stability and reducing the likelihood of alveolar collapse.

The exchange of gases (O_2 & CO_2) between the alveoli & the blood occurs by simple diffusion: O_2 diffusing from the alveoli into the blood & CO_2 from the blood into the alveoli. Diffusion requires a concentration gradient. So, the concentration (or pressure) of O_2 in the alveoli must be kept at a higher level than in the blood & the concentration (or pressure) of CO_2 in the alveoli

---

[1] $\triangle P = 2.T/r$: The pressure inside a bubble exceeds the pressure outside the bubble by twice the surface tension, divided by the radius. In other words, the smaller a bubble, the more the pressure inside it exceeds the pressure on the outside.

must be kept at a lower lever than in the blood. We do this, of course, by breathing - continuously bringing fresh air (with lots of O_2 & little CO_2) into the lungs & the alveoli.

It is a rising concentration of **carbon dioxide** — not a declining concentration of oxygen — that plays the major role in regulating the ventilation of the lungs. As the CO_2 content of the blood rises above normal levels, the pH drops [$CO_2 + H_2O \rightarrow HCO_3^- + H^+$],

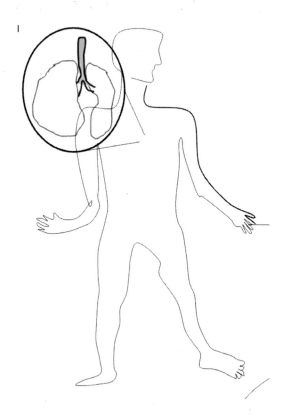

In patients with fair skin, the lungs' size is greater in comparison with patient of dark skin.

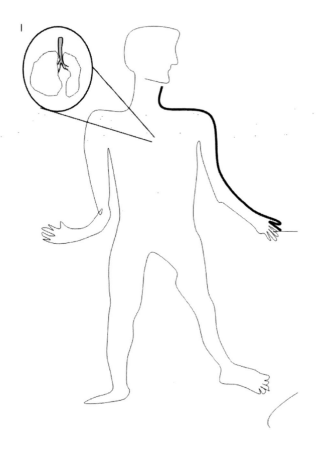

Opposite people with dark skin characteristically have smallest lungs than white people.

Breathing pattern during exercise in young black and Caucasian subjects

F. J. Cerny

Abstract

Lung volumes in sex-, age-, height-, and weight-matched Black subjects are 10-15% lower than those in Caucasians. To determine whether this decreased lung volume affected the ventilatory adaptation to exercise, minute ventilation (VE), its components, frequency (f) and tidal volume (VT), and breathing pattern were observed during incremental cycle-ergometer exercise. Eighteen Caucasian (age 8-30 yr) and 14 Black (age 8-25 yr) subjects were studied. Vital capacity (VC) was lower (P less than 0.001) in the Black subjects [90.6 +/- 8.6 (SD) vs. 112.9 +/- 9.9% predicted], whereas functional residual capacity/total lung capacity was higher (P less than 0.05). VE, mixed expired O2 and CO2, VT, f, and inspiratory

Journal of Applied Physiology
June 1, 1987 vol. 62 no. 6
2220-2223

Racial Differences in Ocular Oxidative Metabolism

Implications for Ocular Disease

Carla J. Siegfried, MD; Ying-Bo Shui, MD; Nancy M. Holekamp, MD; Fang Bai, MD; David C. Beebe, PhD

Results: The PO_2 value was significantly higher in African American patients at all 5 locations compared with Caucasian patients. Adjusting for age increased the significance of this association. Adjusting for race revealed that age was associated with increased PO_2 beneath the central cornea.

Conclusions: Racial differences in oxygen levels in the human eye reflect an important difference in oxidative metabolism in the cornea and lens and may reflect differences in systemic physiologic function. Increased oxygen or oxygen metabolites may increase oxidative stress, cell damage, intraocular pressure, and the risk of developing glaucoma. Oxygen use by the cornea decreases with age.

Arch Ophthalmol. 2011;129(7):849-854

Notices that the authors' conclusion about that pO2 value was significantly higher in African Americans compared with Caucasian patients are based in racial and genetic issues. Our explanation is quite simplest: more melanin more photosynthesis, therefore highest levels of diatomic Oxygen.

The smooth muscle in the walls of the bronchioles is very sensitive to the concentration of carbon dioxide. A rising level of CO_2 causes the bronchioles to dilate. This lowers the resistance in the airways and thus increases the flow of air in and out.

The diffusing capacity for carbon monoxide increases with exercise, and the increase has been thought to be dependent on an increase in the filling of the pulmonary capillary bed, an increase in the number of capillaries that are perfused.

Composition of atmospheric air and expired air in a typical subject. Note that only a fraction of the oxygen inhaled is taken up by the lungs.

Component	Atmospheric Air (%)	Expired Air (%)
N_2 (plus inert gases)	78.62	74.9
O_2	20.85	15.3
CO_2	0.03	3.6
H_2O	0.5	6.2
	100.0%	100.0%

Only in the alveoli does actual gas exchange takes place. There are some 300 million alveoli in two adult lungs. These provide a surface area of some 160 m^2 (almost equal to the singles area of a tennis court and 80 times the area of our skin!). An average adult male can flush his lungs with about 4 liters of air at each breath. This is called the **vital capacity**. Even with maximum expiration, about 1200 ml of **residual air** remain.

The pulmonary circulation is unique in that increased flow is accompanied by the opening up of new channels in the absence of any striking change in pulmonary arterial pressure.

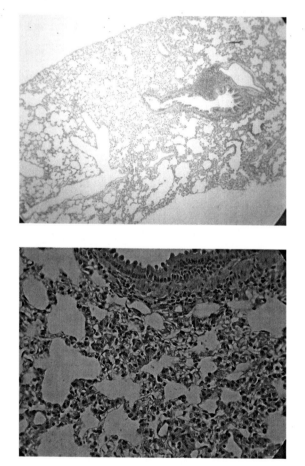

Histological section of a rat´s lung, notice the alveolus tissue.

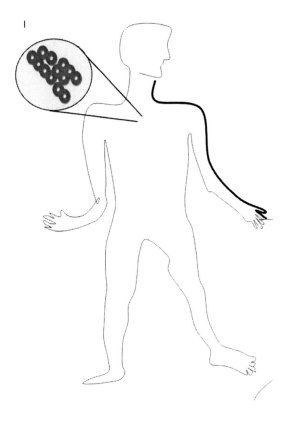

Dark skins people need less amount of hemoglobin, the difference between black and white vary from 1 to 2 g/100 ml.

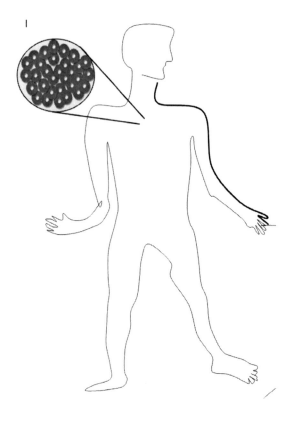

Caucasian patients characteristically have highest level of hemoglobin in comparison with black people.

Lifelong Differences in Hemoglobin Levels Between Blacks and Whites*

STANLEY M. GARN, PH.D., NATHAN J. SMITH, M.D.,
and
DIANE C. CLARK, A.S.,

*Center for Human Growth and Development,
University of Michigan,
and
Department of Pediatrics,
University of Washington School of Medicine*

moglobin levels. Analyzed age by age, from the first year of life through the 9th decade, blacks proved systematically lower than whites by approximately 1gm/100ml.

tests for trend. Thus, in a total of 27,900 subjects, using median values of hemoglobin to eliminate errors due to skewness, there is a systematic difference approximating 1.0gm/100ml. This systematic difference in hemoglobin level is fully evident even during the period of rapid adolescent gain in hemoglobin levels in the male, and during the period of declining hemoglobin levels in the 7th and 8th decades. Black-white differences in hemo-

These differences have been noted since the past century; however remain unexplained until the discovery of the human photosynthesis.

The morphological structure of the lung is unique. The major part of it is "spongy" tissue that allows free exchange of gases between the blood and the atmosphere by arranging for an intimate contact between capillary blood and alveolar air.

The basic element is the alveolus, with a mean diameter of approximately 280 um, its wall is enclosing a dense network of capillaries distributed in the form of continuous hexagons. The mean maximum intercapillary distance in the network is of the order of 9 um. With dimensions of this order of magnitude, all of the water in the tissue phase would be expected to be available for diffusion equilibration with the water in perfused capillaries.

Human Photosynthesis and the Lung Function

The lung is a not well understood organ, for instance through the pulmonary artery, the level of Oxygen is very low and the CO_2 level is high. Oppositely the pulmonary vein brings blood with a pO2 of 98 % and CO_2 level of 20 %. The question is the lung´s cells in what moment absorbs oxygen?

If we take into account the extraordinary capacity of every single cell of the body to split and reform the water molecule and therefore releases diatomic oxygen and diatomic hydrogen, then the answer begin to be coherent. Human´s lung is an organ with a high turnover rate of water dissociation and reformation (human photosynthesis), and its high water content, about 80 %[12]; is a very good proof.

Therefore we could say that lung´s tissue produces diatomic oxygen at their own by mean of the intrinsic property of melanin to dissociate and reform the water molecule, processes that are not symmetrical in time.

The remarkable speed of adaptation of newborn from the placenta to lung as the organ of gas interchange (O_2 and CO_2) and the efficiency of the respiratory apparatus of the newborn infant continue to awe those who ponder the adjustment. Moreover there is a significant balance between airway dimensions and alveolar surface area, which changes with growth of the lung and the rest of the body. In the first days of life the respiratory quotient falls, body weight may fall about 10 %, and in the ensuing 5/6 months weight doubles. During this period changes in the circulation are underway as the fetal channels, foramen ovale and ductus arteriosus close, and pulmonary vascular resistance falls. It is difficult enough to comprehend the subtleties of ventilation-perfusion relations in the adult lung< their changing patterns during the cardiopulmonary adaptation at birth and shortly thereafter is even more complex.

The description of respiration in the newborn period requires measurements in the first minutes of life, followed by their repetition in a matter of hours and also days. A further requirement is methodology suitable miniaturized to permit examination of premature as well as term newborn infants.

The fetal lung contains fluid, which is in part derived from the lung itself and in part may be aspirated amniotic fluid. The probability of a finite volume of fluid in the lung at birth was deduced by pathologists on the basis of observation of the distention of sequestered portions of lung and occasional squamous cells from the fetal skin seen even in lungs of infants who died from non-pulmonary causes.

Studies of animal lungs show their greater weight in the fetal state compared with that after the initiation of respiration, and the lower specific gravity of fetal compared with gas-free newborn lung. Fluid can be seen flowing from the trachea of the exteriorized fetal animal and analysis of the fluid suggests the theory that it is an ultra filtrate of fetal blood, like kidney, however it is more precise to say that the Solis-Herrera cycle is skewed to the left. The volume of fluid normally present in the fetal lung is estimated at approximately one-half the functional residual capacity. The rate of its disappearance is assumed to be very rapid on the basis of the establishment of gaseous FRC with the first breath and the loss of lung weight within ten minutes of the initiation of respiration.

In a weight basis adult lungs are approximately 18 times as heavy as infant lungs. However, adult lungs are 33 times as compliant. Thus, expressed per kilogram of lung tissue, infant lungs would appear to be less compliant tan those of the adult. However, using functional residual volume as a basis of comparison, adult lungs are similar in compliance to those of infants (compliance/FRCratio=0.065 in infants and 0.063 in adults).

The resistance to air flow in the infant's lungs is only 15 times that of the adult, considerably less than would be expected if infant air passages were reduced in size and length in proportion to weight. The average pulmonary work during quiet respiration estimated for the hypothetical infant of 3 Kg is approximately 1 per cent of the total basal metabolism, as in adult, if the efficiency of the respiratory muscles is assumed to be between 5 and 10 % for both age groups. In uterus, fetus has a breathing rate of 20 per hour. The theoretical minimum of work of respiration in the normal newborn infant occurs at approximately 37 respirations per minute.

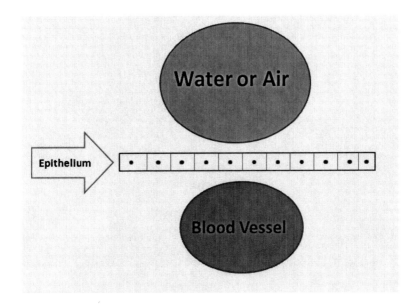

Histological characteristics of the lung or gill are very similar; therefore we could think that fetus´ lung works as gill does during intrauterine life.

In my opinion, the main aim of the fish gills and/or mammal lungs is the interchange of CO_2 due to its toxic effects, more than the oxygen transportation from the outer to inner side because the organism is able to produce diatomic oxygen at its own. Even the striking case of convergent evolution of the carbonic anhydrase enzyme, that speed up ten thousand times the velocity of reaction of combination of CO_2 with water in both ways shows the importance of the driving of CO_2 in the mammal metabolism.

Carbonic Anhydrase

$$CO_2 + H_2O \leftrightarrow H_2CO_3$$

For the maintenance of life, theoretically fishes utilize the dissolved oxygen in the water that is relatively scarce and is very doubtful that is enough to whales. By means of the gills, the blood, though separate by a delicate layer of epithelium, is brought into close contact with the water. Through this thin covering supposedly oxygen is absorbed by the blood, and carbon dioxide and other waste products are given off in a comparable manner to the respiratory exchange of air breathing animals. However the amount of dissolved diatomic oxygen present in water is little in order to gives support to the life of fishes of different size and weight. Perhaps in minute fishes dissolved oxygen could be enough, but after certain size, volume and weight, water dissolved oxygen is clearly not sufficient to sustain a metabolic system based in glucose oxidation as source of energy, instead if we take into account the unsuspected capacity of melanin to split and reform the water molecule as main source of energy, then the explanation sounds more coherent and the ever presence of melanin in the fish is also explained.

The lungs of fetus have a respiration rate of already 20 by hour, a similar rate as in fishes[13].

The rate and depth of breathing in normal adults are adjusted with the result that alveolar ventilation is accomplished with the minimum expenditure of total respiratory work.

CHAPTER 16

GENES AND HUMAN PHOTOSYNTHESIS

Genes were after photosynthesis.

Mendel defined the basic nature of the gene as a particulate factor that passes unchanged from parent to progeny. A gene may exist in alternative forms (alleles). In diploid organisms, which have two sets of chromosomes, one copy of each chromosome is inherited from each parent. The chromosomes carry the genes.

Each chromosome consists of a linear array of genes. On the genetic maps of higher organisms established during the first half of the past century, the genes are arranged like beads on a string. The gene is to all intents and purposes a mysterious object (the bead) whose relationship to its surroundings (the string) is unclear.

The genetic material of a chromosome consists of an uninterrupted length of DNA that represents many genes. A genome consists of the entire set of chromosomes for any particular organism, and therefore comprises a series of DNA molecules, each of which contains a series of many genes. The major component in prokaryote and eukaryote chromosomes is DNA (deoxyribonucleic acid). The genetic material of all known organisms and many viruses is DNA. The general principle of the nature of the genetic material, then, is that it is always nucleic acid; in fact, it is DNA except in the RNA viruses.

A nucleic acid consists of a polynucleotide chain. The backbone of the chain consists of an alternating series of pentose (sugar) and phosphate residues. A purine or pyrimidine nitrogenous base is linked to the sugar. DNA takes its name from its sugar (2 deoxyribose), RNA is named for its sugar (ribose). The difference is that the sugar in RNA has an OH group at the 2 position of the pentose ring.

Each nucleic acid contains four types of base. The same two purines, adenine and guanine, are present in both DNA and RNA. The two pyrimidines in DNA are cytosine and thymine, in RNA uracil is found instead of thymine. The only difference between uracil and thymine is the presence of a methyl substituent at position C_5. The bases are usually referred to by their initial letters, DNA contains A, G, C, T, while RNA contains A, G, C, U.

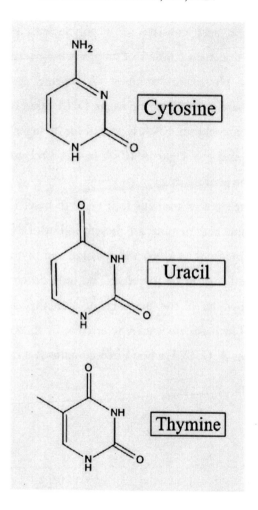

The observation that the bases are present in different amounts in the DNAs of different species led to the concept that the sequence of bases is the form in which genetic information is carried. DNA has the form of a regular helix, making a complete turn every 34 Å (3.4 nm), with a diameter of ˜ 20 Å (2.0 nm). Since the distance between adjacent nucleotides is 3.4 Å, there must be 10 nucleotides per turn.

The density of DNA suggests that the helix must contain two polynucleotide chains. The constant diameter of the helix can be explained if the bases in each chain face inward and are restricted so that a purine is always opposite a pyrimidine, avoiding partnerships of purine-purine (too thick) or pyrimidine-pyrimidine (too thin).

Irrespective of the actual amount of each base, the proportion of G is always the same as the proportion of C in DNA, and the proportion of A is always the same as that of T. Watson and Crick proposed that the two polynucleotide chains in the double helix associate by hydrogen bonding between the nitrogenous bases. G can hydrogen bond specifically only with C, while A can bond specifically only with T. These reactions are described as base pairing, and the paired bases (G with C and A with T) are said to be complementary.

The model requires the two polynucleotide chains to run in opposite directions (antiparallel) as illustrated in the above figure. Looking along the helix, therefore; one strand runs in the 5´→3´direction, while its partner runs 3´→5´.

The sugar phosphate backbone is on the outside and carries negative charges on the phosphate groups. When DNA is in solution in vitro, the charges are neutralized y the binding of metal ions, typically by Na^+. In the natural state in vivo, positively charged proteins provide some of the neutralizing force. These proteins (and water also) play an important role in determining the organization of DNA in the cell.

Proceeding along the helix, bases lie on inside and are stacked above one another, in a sense like a pile of plates. Each base pair is rotated ~ 36° around the axis of the helix, relative to the next base pair. So ~ 10 base pairs make a complete turn of 360 °. The twisting of the two strands around one another forms a double helix with a narrow groove (~22 Å across) and a wide groove (~22 Å across). The double helix is right handed; the turns run clockwise looking along the helical axis and are named as the B-form of DNA.

DNA replication:

It is crucial that the genetic material is reproduced accurately. Because the two polynucleotide strands are joined only by hydrogen bonds and immersed in water all the time, they are able to separate without requiring breakage of covalent bonds. The specificity of the base pairing suggests that each of the separated parental strands could act as a template for the synthesis of a complementary daughter strand. The principle is that a new daughter strand is assembled on each parental strand. The sequence of the daughter stand is dictated by the parental strand; an A in the parental strand causes a T to be placed in the daughter strand, a parental G directs incorporation of a daughter C, and so on.

Each of the daughter duplexes is identical in sequence with the original parent and contains one parental strand and one newly synthesized strand. This behavior is called semi-conservative replication. The disruption of the structure is only transient and is reversed as the daughter duplex is formed. So only a small part of the DNA loses the duplex structure at any moment.

The double helical structure is disrupted at the junction between two regions, called the replication fork. We must take in account that all the time the entire chromosome and therefore the genes and their DNA is immersed in water, where the water tend to dissolve constantly any compound present in it. So any part of the structure requires energy in order to keep the shape besides any change, in spite can be conformational or chemical, and requires energy also.

The synthesis of nucleic acids is catalyzed, an activity that requires energy indeed; by specific enzymes that are also immersed in water, which recognize the template, and requires energy to stand still; so can undertake the task of catalyzing the addition of subunits to the polynucleotide chain that is synthesized. The enzymes are named according to the type of chain that is synthesized: DNA polymerases synthesize DNA, and RNA polymerases synthesize RNA. However are not specified where the energy required for it comes from.

In this moment let us to recall generalities of the cell structure:

Briefly, the basic properties of cells are: Cells are highly complex and organized, cells possess a genetic program and the means to use it; cells are capable of producing more of them; cells carry out a variety of chemical reactions; cells engage in numerous mechanical activities; cells are able to respond to stimuli; cells are capable of self-regulation; cells are immersed in water; and more important: cells are able to acquire and utilize energy.

The developing and maintenance of the complexity requires the constant input of energy[14]. Virtually and mainly all the energy required by life on Earth´s surface arrives in the form of electromagnetic radiation from the sun. The energy of light is trapped by light-absorbing pigments present in the photosynthetic cells. And melanin "pigment" is not an exception. Traditionally is believed that light energy is converted by photosynthesis into chemical energy that is "stored" (energy cannot be stored) in energy-rich carbohydrates, such as sucrose and starch, but more exactly we could say biomass-rich molecules. Wrongly is believed that the energy trapped in these molecules during photosynthesis in plants provides the fuel that runs the activities of nearly all the organisms on Earth.

A wrongly belief is that for most animal cells, energy arrives prepackaged, usually in the form of the sugar glucose. However glucose has not energy to practical aims; therefore is just a source of biomass, the better indeed; but is a biomass source only. In humans, glycolysis is performed by the same enzymes that are present in all living cells, exactly the same enzymatic machinery. In the liver we have glucokinase and hexokinase; and the glucose levels in the blood are maintained within certain ranges by the release of this molecule by the liver, beside other available mechanisms. Thus the cell is able to introduce glucose into her, and once in a cell, the glucose is disassembled in such a way that its biomass available, not energy; can be stored in an available form, usually as ATP. But the cell´s energy requirements are determined by the running all the myriad energy-requiring activities in a cell. Energy cannot be stored in anyway instead the building blocks of the biomass can really be stored.

Characteristics that distinguish prokaryotic and eukaryotic cells

Features held in common by the two types of cells:

1. Plasma membrane of similar construction

2. Genetic information encoded in DNA using identical genetic code

3. Similar mechanism for transcription and translation of genetic information, including similar ribosomes

4. Shared metabolic pathways (for instance glycolysis and TCA cycle)

5. Similar apparatus for conservation of chemical energy as ATP, located in the plasma membrane of prokaryotes and the mitochondrial membrane in eukaryotes. But in our opinion, and at light of the discovery of the intrinsic property of melanin to split and reform the water molecule, this paragraph must be rewritten in the next way: Similar apparatus for the transformation of photonic energy into chemical energy, so then, ATP molecule is a very important metabolic intermediate and not the universal currency of energy, instead ATP is more the universal currency of phosphate molecules, compounds that have significant role in the biomass processes.

6. Similar mechanisms of photosynthesis, chlorophyll for cyanobacteria and green plants, and melanin for nearly all living cells.

7. Similar mechanisms for synthesizing and inserting membrane proteins

8. Proteasomes of similar construction

9. In both the universal solvent is water.

In regards the features of eukaryotic cells not found in prokaryotic:

1. Division of cells into nucleus and cytoplasm, separated by a nuclear envelope containing complex pore structures. The genetic material of a prokaryotic cell is present in a nucleoid; a poorly demarcated region of the cell that lacks a boundary to separate it from the surrounding cytoplasm. In contrast, eukaryotic cell possess a nucleus; a region bounded by a complex membranous structure called the nuclear envelope. This difference in nuclear structure is the basis for the terms prokaryotic (pro=before, karyon=nucleus) and eukaryotic (eu=true, karyon=nucleus).

    Prokaryotic cells contain relatively small amounts of DNA; the total length of the DNA of a bacterium ranges about 0.25 mm to about 3 mm, which is sufficient to encode between several hundred and several thousand proteins.

    Although the simplest eukaryotic cells have only slightly more DNA (4.6 mm in yeast encoding about 6200 proteins) than the most complex prokaryotes, most eukaryotic cells contain an order of magnitude more genetic information. Both prokaryotic and eukaryotic cells have DNA-containing chromosomes.

Each of the numerous chromosomes of a eukaryotic cell contains a linear molecule of DNA associated tightly with protein, while the single circular chromosome of a prokaryotic cell consists essentially of naked DNA. Considering its importance in the storage and utilization of genetic information, the nucleus of a eukaryotic cell has a rather undistinguished morphology. The contents of the nucleus dissolved in water; are present as viscous, amorphous mass of material enclosed by a complex nuclear envelope that forms a boundary between the nucleus and cytoplasm. Included are the chromosomes, which are present as highly extended nucleoproteins fibers, termed chromatin; the nuclear matrix, which is a protein-containing fibrillar network; one or more nucleoli, which are irregular shaped electron dense structures that function in the synthesis of ribosomal RNA and the assembly of ribosomes; and the nucleoplasm, the fluid substance, water base; in which the solutes of the nucleus are dissolved. Given the absence of mitochondria and ATP in the cell greatest organelle: the nucleus, then the necessary energy for their functions, from where?

The nuclear envelope consists of several distinct components. The core of the nuclear envelope consists of two cellular membranes arranged parallel to one another and separated by an inter-membrane space of 10 to 15 nm. The membranes of the nuclear envelope serve as a barrier that keeps ions, solutes, ATP and macromolecules from passing between the nucleus and cytoplasm. The two membranes are fused at sites forming circular pores that contain complex assemblies of proteins. The average mammalian cell contains approximately 3000 pores. The outer membrane is generally studded with ribosomes and is occasionally seen to be continuous with membrane of the rough endoplasmic reticulum.

The nuclear envelope is the barrier between the nucleus and cytoplasm, and nuclear pores are the gateway across the barrier. The nuclear envelope is a hub of activity for the movement for the movement of RNAs and cytoplasm. The replication and transcription of genetic material within the nucleus require the participation of large number of proteins that are synthesized in the cytoplasm and transported across the nuclear envelope. Conversely, the mRNAs, tRNAs, and ribosomal subunits that are manufactured in the nucleus must be transported through the nuclear envelope in the opposite direction. Some components, such as snRNAs, move in both directions; they are synthesized in the nucleus, assembled into RPN particles in the cytoplasm, and then shipped back to the nucleus where they function in mRNA processing. To appreciate the magnitude of the traffic between the two major cellular components, consider a HeLa cell, which is estimated to contain about 10,000,000 ribosomes.

To support its growth, a single HeLa cell nucleus must import approximately 560,000 ribosomal proteins and export approximately 14,000 each minute. It is a huge activity, the undoubtedly requires energy; but without mitochondria or ATP, the energy for the nucleus from where?

By other side, an average human cell contains about 6 billion base pairs of DNA divided among 46 chromosomes, the value for a diploid, unreplicated number of chromosomes. Each unreplicated chromosomes contains a single, continuous DNA molecule; the larger the chromosome, the longer the DNA it contains. Given that each base pair is about 0.34 nm in length, 6 billion base pairs would constitute a DNA molecule fully 2 m long. In addition, the DNA in a cell binds a large amount of water, about six water molecules per base pair; therefore the constant water transportation requires energy also because it is not stagnant water, instead the water is in continuous movement; approximately 100 times the volume of water in a cell crosses the plasma membrane every second[15].

How is it possible to fit 2 meters of hydrated DNA into a nucleus only 10 µm in diameter and, at the same time, maintain the DNA in a state that is accessible to enzymes and regulatory proteins? Just as important, how is the single DNA molecule of each chromosome organized so that it does not become hopelessly tangled with the molecules of other chromosomes? The answer lays in an adequate and constant energy supply and thereafter a very fine tuning expenditure of it developed along four billion years of evolution; for instance the remarkable complex manner in which the DNA molecule is packaged, a set of sophisticated processes that requires indeed also energy.

It has long been known that chromosomes are composed of DNA and associated protein, which together are called chromatin. There are two types of proteins in chromatin: histones and nonhistone chromosomal proteins. Histones are small, well-defined basic proteins, whereas nonhistone chromosomal proteins include a large number of widely diverse structural enzymatic, and regulatory proteins, most of which have yet to be characterized. The synthesis, maintenance, function, degradation and replacement of chromatin have a very first requirement in common: energy.

The orderly packaging of eukaryotic DNA depends of histones, therefore this mentioned sequential order, and the DNA itself and histones also, requires and even more: is completely depending of a constant source of energy. Many key processes that occur within the nucleus, including transcription, RNA processing, and replication, are thought to be compartmentalized. Any processes, which can be or not key, have a very first requirement in common: energy. Notice that the nucleus and their structures are imbibed in water all the time, night and day, so even to keep the shape, cell nucleus requires more energy. Life expression seems a delicate dynamic balance between energy and biomass. For Life available energy in a constant manner is essential but biomass is different; certain building blocks must be present almost all the time, for instance glucose, because the cell is able to split, to cut, or extend it, modifying the length, and its branching and composition by means of addition of atoms or molecules composed by carbon, nitrogen, oxygen or hydrogen.

Vertebrates are composed of hundreds of different cell types, each far more complex than a bacterial and each requiring a distinct battery of proteins that enable it to carry out specialized activities.

Both types of energy released by melanin: diatomic hydrogen when the water molecule is splinted and a brief pulses of high energy electrons when hydrogen and oxygen are recombined is very useful for the cell. For instance, the cell fusion in Dolly case was accomplished by bringing the two types of cells into contact and subjecting them to a brief electric pulse, which also served to stimulate the egg to begin embryonic development.

Because of the tremendous amount of DNA found in a eukaryotic cell and the large number of different proteins being assembled (already 100,000 different polypeptides during their lifetime); regulating eukaryotic gene expression is an extraordinarily complex process that is not yet understood.

For instance, hemoglobin accounts for more than 95 % of the protein in a red cell, yet the genes that code for the hemoglobin polypeptides represents less than one-millionth of its total DNA. Not only does the cell have to find this genetic needle in the chromosomal haystack, it has to regulate its expression to such a high degree that production of these few polypeptides becomes the dominant synthetic activity of the cell. Because the chain of events leading to the synthesis of a particular protein consists of a number of discrete steps; there are several levels at which control might be exercised, even at extracellular level, as epigenetic concept define; however, in spite of their complexity, gene expression, gene regulation; gene maintenance; etc., have a very unavoidable first requirement: energy.

But nucleus cell has not ATP or mitochondria, so the energy from where? The answer is from melanosomes, that contain the melanin granules and characteristically surrounding the nucleus at level of per-nuclear space, and at that level, melanin granules release the energy symmetrically and in all directions, like growing spheres of diatomic both hydrogen and oxygen, alternating with spheres of reformed water and therefore a flow of electrons of high energy. This growing spheres coalescing at the center of the cell nucleus, forming a high energy zone that is able to support the huge energy requirements of the greater structure of the cell: the nucleus.

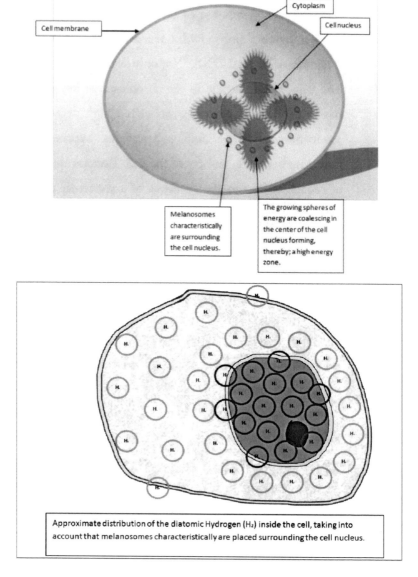

Cytoplasm

Cell membrane

Cell nucleus

Melanosomes characteristically are surrounding the cell nucleus.

The growing spheres of energy are coalescing in the center of the cell nucleus forming, thereby; a high energy zone.

Approximate distribution of the diatomic Hydrogen (H_2) inside the cell, taking into account that melanosomes characteristically are placed surrounding the cell nucleus.

In my opinion, the understanding and treatment of the Gene Diseases must change. The very first requirement of a particular gen and therefore the entire genome in order to expressing and makes their functions adequately is energy, biomass is secondarily; and this crucial energy supply to the nucleus is provided by melanin. We must keep in mind that cell nucleus has neither mitochondria nor ATP. Those "families" affected by something classified as hereditary disease, perhaps is just a group of people exposed to similar toxic compounds, for instance herbicides or toxic industrial compounds that sooner or later reach the drinking water. For the body the name of the disease is not important and human classifications neither.

Possibly ATP is just a metabolic intermediate, for instance, as phosphate regulator. When the cell needs to "activate" some substance, in other words: to solubilize it in certain manner, because this "solubilization" process can take place in several ways; for instance, a phosphate group is attached; and once it is phosphorilated, the substance cannot get away from the cell.

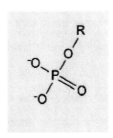

A phosphate group is composed by a phosphorous attached to four oxygen atoms (PO_4) (phos=light; phor= bear, carry); the element phosphorus was so named because some forms "glows in the dark".

Phosphorylation is defined as the addition of a phosphate group, such as PO_3H_2 from ATP to a specific intracellular protein such as an enzyme important in a particular metabolic pathway. Phosphorylation causes the protein to change its shape and function (either activating or inhibiting it). This altered protein brings about the desired function. For example, a given ion channel may be opened or closed or the activity of a particular enzymatic protein that regulates a specific metabolic event may be increased or decreased. Intracellular enzymes inactivate the other participating chemicals so that the response is terminated. Otherwise, once triggered the response would go on indefinitely until the cell ran out of necessary supplies.

Phosphate groups have a very important role in the ordering of the sequence of intracellular events that ultimate controls a particular cellular activity important in the maintenance of homeostasis (steady state) such as membrane transport, secretion, metabolism or contraction. Phosphate groups appear to modify the energetic level of those compounds that eventually are combined with them. However these energetic level's modifications are not due that ATP, GTP or the phosphate group itself is source of energy, in reality the maintenance of the constant supply of energy to the whole cell necessary for the myriad of cellular functions comes from human photosynthesis process. The modifications in the energetic levels of compounds relative to phosphate groups undergo to slightly changes that are enough to activate or inhibit mainly.

Compounds containing phosphorus play many critical rolls in biochemistry. If organic chemistry were only taught for life scientists, phosphorous compounds would receive a great deal more attention. A short list of examples of significant biological compounds with phosphorus would include both nucleic acids, in which nucleosides are connected by phosphate linkages, and phospholipids, the major component of biological membranes.

Also Inorganic phosphate (Pi) is required for cellular function and skeletal mineralization. Serum Pi level is maintained within a narrow range through a complex interplay between intestinal absorption, exchange with intracellular and bone storage pools, and renal tubular reabsorption. Pi is abundant in the diet, and intestinal absorption of Pi is efficient and minimally regulated. The kidney is a major regulator of Pi homeostasis and can increase or decrease its Pi reabsorptive capacity to accommodate Pi need. The crucial regulated step in Pi homeostasis is the transport of Pi across the renal proximal tubule. Type II sodium-dependent phosphate (Na/Pi) co-transporter (NPT2) is the major molecule in the renal proximal tubule and is regulated by hormones and non-hormonal factors[16]. We could say that one of the main phosphates´ functions is as metabolic regulator.

Summarizing: cell´s nucleus depends completely of the energy released through out by the humans photosynthesis system composed by Light/ Melanin/ Water; arranged in order of abundance in nature.

Endoplasmic Reticulum:

The endoplasmic reticulum (ER) is an elaborate fluid-filled membranous system distributed extensively throughout the cytosol, therefore is immersed in water. It is primarily a protein – and lipid- manufacturing factory. Two distinct types of endoplasmic reticulum –the smooth ER and the rough ER- can be distinguished. The smooth ER is a meshwork of tiny interconnected tubules, whereas the rough ER projects outward from the smooth as stacks of relatively flattened sacs. Even though these two regions differ considerably in appearance and function, they are continuous with each other. In other words, the ER is one continuous organelle with many interconnected channels. The relative amount of smooth and rough varies between cells, depending on the activity of the cell.

The outer surface of the rough ER membrane is studded with small; dark-staining that gives it a rough or granular appearance under a light microscope. These particles are ribosomes, which are ribosomal RNA-protein complexes, which synthesize proteins under the direction of the nuclear DNA. Messenger RNA carries the genetic message from the nucleus to the ribosome workbench, where protein synthesis takes place. Not all ribosomes in the cell are attached to the rough ER. Unattached or free ribosomes are dispersed throughout the cytosol.

The rough ER, in association with its ribosomes, synthesizes and releases a variety of new proteins into the ER lumen, the fluid-filled space enclosed by the ER membrane. Both rough and smooth ER is immersed in water all the time. The proteins synthesized by the rough ER serve one of two purposes:

1. Some proteins are destined for export to the cell´s exterior as secretory products, such as protein hormones and enzymes. (All enzymes are proteins).

2. Other proteins are transported to sites within the cell for use in constructing new cellular membranes, either new cytoplasm membrane or new organelle membrane; or other protein components of organelles.

The Smooth ER

The smooth ER does not contain ribosomes, so it is "smooth". Lacking ribosomes, it is not involved in protein synthesis. Instead, it serves a variety of other purposes that vary to different cell types.

In the majority of cells, the smooth ER is rather sparse and serves primarily as a central packaging and discharge site for molecules that are to be transported from the ER. Newly synthesized proteins and lipids pass from rough ER to gather in the smooth ER. Portions of the smooth ER then "bud off", forming transport vesicles that contain the new molecules enclosed in a spherical capsule of membrane derived from the smooth ER. Newly synthesized membrane components are rapidly incorporated into the ER membrane itself to replace the membrane that was used to wrap up the molecules in the transport vesicle. Transport vesicles move to the Golgi complex, described in the next section for further processing of their cargo.

The smooth ER is abundant in cells that specialize in lipid metabolism, for instance, cells that secrete lipid-derived steroid hormones. The membranous wall of the smooth ER, like of the rough ER, contains enzymes for the synthesis of lipids. The lipid-producing enzymes in the membranous wall of the rough ER alone are insufficient to carry out the extensive lipid synthesis necessary to maintain adequate steroid-hormone secretion levels. These cells have an expanded smooth-ER compartment to house the additional enzymes necessary to keep pace with demands for hormone secretion. Both smooth ER and their enzymes are immersed in water.

In liver cells, the smooth ER has a special capability. It contains enzymes that are involved in detoxifying harmful substances produced within the body by metabolism or substances that enter the body from the outside in the form of drugs or other foreign compounds. These detoxification enzymes alter toxic substances so that the latter can be eliminated more readily in the urine. The amount of ER available in the liver cells for the task of detoxification can vary dramatically, depending on the need. The liver and its content are immersed in water all the time.

Muscle cells have an elaborate, modified smooth ER known as the sarcoplasmic reticulum, which stores Calcium that plays an important role in the process of muscle contraction.

Interestingly, neither rough nor smooth ER has mitochondria or ATP into its lumen; so the huge amounts of energy required by the protein synthesis processes and packaging, from where? Furthermore, all enzymatic activity requires activation energy, otherwise the chemical process cannot be initiated, but thereafter energy is also required during in and at the end of the process, otherwise the end process cannot be finished adequately. So a constant requirement in any intra smooth or rough ER process is energy. And this essential supply of energy to the smooth and rough ER; from where?

The answer again lays in the intrinsic property of melanin to split and reform the water molecule. Melanosomes, characteristically, are place in the per-nuclear space, thereby surrounding the nucleus of all living cells, forming a shell-like. By other side, the rough ER covers almost completely the cell´s nucleus so the organelle is able to trap the molecular Hydrogen of the growing spheres of gas and liquid those, in alternating way; are released constantly by the granules of melanin contained in the melanosomes. Therefore; the disposition of the rough ER around the cell´s nucleus is a form to ensure that the great part of these two types of energy are cached, given the enormous amount of energy required for the protein synthesis activity of the rough and smooth ER. Diatomic Hydrogen does not requires active transport or even specialized pores, because is the smallest atom and easily can pass through the cell membrane of the rough and smooth ER; or any other organelle; due to, thanks that molecular Hydrogen follows the rules of simple diffusion, tending, therefore; to diffuse passively throughout the entire cytoplasm, accomplishing, at the same time, numerous crucial functions as antioxidant, the releasing of the carried energy adequately; and so on.

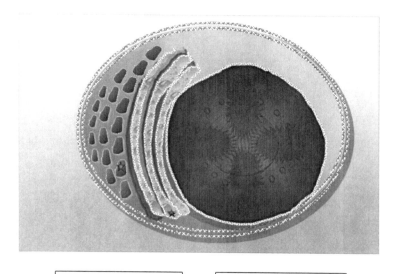

| Smooth Endoplasmic Reticulum ★ | Rough Endoplasmic Reticulum ★ |

The cell´s nucleus with melanosomes and per-nuclear space that is surrounded almost completely by the rough endoplasmic reticulum.

Intracellular organelles are disposed in a way that meanwhile more energy needs, more closely will be to the cell nucleus area.

ATP and therefore glucose cannot be the source of energy to the rough and smooth ER, because ATP is found in the cytoplasm and into mitochondria, far away, relatively; from smooth and rough ER, and more important, each ATP molecule must be reconstituted each twenty seconds or less, so it is not enough time to the ATP molecule to pass back and forward through the membrane of ER.

It is very important that diatomic Hydrogen does not combine with water, therefore it travel "swimming" through the entire cytoplasm and membranes easily, carrying the crucial energy and distributing it along the cytoplasm elements as organelles, and the wide variety of intracellular compounds; combining with and modifying them all in one or other way and in unceasingly manner practically the complete cell content, and the water is not an exception.

Golgi complex

Closely associated with the endoplasmic reticulum is the Golgi complex. Each Golgi complex consists of a stack of flattened, slightly curved membrane-enclosed sacs, or cisternae. The sacs within each Golgi stack are not physically connected Note that the flattened sacs are thin in the middle but have dilated, or bulging edges. The number of Golgi complexes varies, depending on the cell type. Some cells have only one Golgi stack, whereas cells highly specialized for protein secretion may have hundreds of stacks.

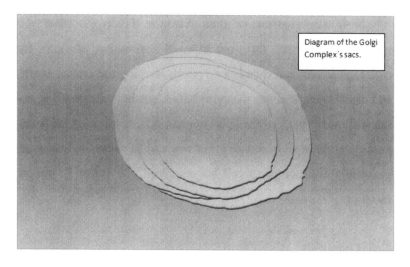

Diagram of the Golgi Complex´s sacs.

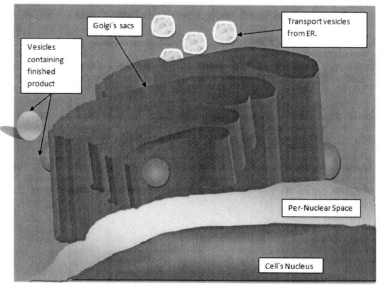

Golgi´s sacs

Transport vesicles from ER.

Vesicles containing finished product

Per-Nuclear Space

Cell´s Nucleus

The majority of the newly synthesized molecules that have just budded off from the smooth ER enter a Golgi stack. When a transport vesicle carrying its newly synthesized cargo reaches a Golgi stack, the vesicle membrane fuses with the membrane of the sac closest to the center of the cell. The vesicle membrane opens up and becomes integrated into the Golgi membrane, and the content of the vesicle are released to the interior of the sac. We must keep in mind that Golgi sacs and transport vesicles are immersed in water and have water inside all the time requiring energy not only to perform the synthesis and maturation processes but also to keep the shape because water tend to dissolve everything. By other side, any change, displacement, traffic, membranes that open up, fuse or close are energy-requiring processes absolutely. No one chemical or biochemical reaction can take place if energy is not available constantly.

During the transit throughout Golgi complex, at least two important interrelated functions take place:

1. Processing the raw materials into finished products. Within the Golgi complex, the "raw" proteins from the ER are modified into their final form, largely by adjustments made in the sugars attached to the protein; therefore energy and biomass is required, indeed. The biochemical pathways that the proteins undergo during their passage through the Golgi complex are elaborate, precisely programmed and specific for each final product. Every single process cited above requires available energy besides building blocks.

2. Sorting and directing the finished products to their final destination. The Golgi complex is responsible for sorting and segregating different types of products (1) to be secreted to the cell's exterior, (2) to be used for

construction of new plasma membrane, or (3) to be incorporated into other organelles, especially lysosome. Recall that the sorting and segregation are based in biochemical reactions that have in common a very first requirement: energy; and the diatomic hydrogen by one side and the flow of high energy electrons by the other is the better way to impel those processes.

Mitochondria

A single cell may contain as few as hundred or as many as several thousand mitochondria. In contrast plant´s cell contain a fixed number: 250 mitochondria. Mitochondria are considered so far as the energy organelles, or "power plants" of the cell; but what is the source of energy of the mitochondria itself?

Theoretically they extract energy from the nutrients in food and transform it into a usable form for cellular activities. In accordance, Mitochondria extract energy oxidizing glucose or more precisely pyruvate, therefore diatomic oxygen requirements must be liters and liters by second, however, strikingly, our body, through the lungs; just absorbs already 25 % of the available oxygen in alveoli in each breathing movement.

Mitochondria are rod-shaped or oval structures about the size of bacteria. Mitochondria possess their own DNA, distinct from the DNA housed in the cell´s nucleus. Mitochondrial DNA contains the genetic codes for producing many of the molecules the mitochondria needs to generate energy. In my opinion this is a mistake, mitochondria cannot generate energy, and mitochondrion is not autonomous; in other words this structure is not able to generate enough energy even for itself, so mitochondria have not an output of enough energy to the whole cell.

Prominent among the mitochondrial diseases are those that become debilitating in later life, such as some forms of degenerative nervous-system and muscle diseases.

Each mitochondrion is enclosed by a double membrane – a smooth outer membrane that surrounds the mitochondrion itself, and an inner membrane that forms a series of infoldings or shelves called cristae, which project into an inner cavity filled with a gel-like solution known as the matrix. These cristae contain the electron transport proteins. The matrix consists of a concentrated mixture of hundreds of different dissolved enzymes (the citric acid cycle) that prepare nutrient molecules for the final extraction of usable energy (processed of biomass more exactly) by the cristae proteins.

Food can be thought of as the "crude fuel" whereas the ATP is the "refined fuel" for operating the body´s machinery. In my opinion, this sentence must be corrected as follows: Food can be thought as the raw building material whereas ATP is part of the processes that produced the refined building blocks for the formation, implementation and replenishment of biomass´ body.

Dietary food is digested, or broken down, by the digestive system into smaller absorbable units that can be transferred from the digestive tract lumen into the blood. For instance, dietary carbohydrates are broken down primarily into glucose, which is absorbed into the blood. No usable energy is released during the digestion of food, instead energy is required and by other side it is not possible to extract more energy of any system that previously it has. Supposing by one moment that glucose is a source of energy, then the measuring of the ATP molecules finally produced must include the energy used during the digestion of food.

When delivered to the cells by the blood (the transportation through the blood stream is a highly complex process that requires energy), then the nutrient molecules are transported across the plasma membrane into the cytosol (more energy required, indeed). Once glucose or nutrients have reached the cytoplasm cell, ATP is generated from the sequential dismantling (all of them need energy to happen) of absorbed nutrient molecules in three different steps:

- Glycolysis
- The Citric Acid Cycle
- The Electron Transport Chain

Glycolysis:

Among the thousands of enzymes within the cytosol are those responsible for glycolysis, described as chemical process that involve 10 separate sequential reactions that break down the simple six-carbon sugar, glucose into two pyruvic acid molecules, each of which contains three carbons. During glycolysis just part of the energy from the glucose chemical bonds (ca. 50 %) is used to convert ATP into ADP.

Therefore, glycolysis is not very efficient in terms of energy extraction (we cannot get energy where doesn´t be); the net yield is only two molecules of ATP per glucose molecule processed. Much of the energy originally contained in the glucose molecule is still locked in the chemical bonds of the pyruvic acid molecules. The low energy yield of glycolysis is insufficient to support the body´s demand for ATP. This seems as where the mitochondria come into play.

Citric Acid Cycle

The pyruvic acid produce by glycolysis in the cytosol can be selectively transported into the mitochondrial matrix, requiring undoubtedly more energy. Inside the mitochondria, pyruvic acid is further broken down into two carbon-molecules, acetic acid; by enzymatic removal of one of the carbons in the form of carbon dioxide (CO_2); which eventually is eliminated from the body as an end product or waste. CO_2 is the most oxidized form of carbon.

During this breakdown process, a carbon-hydrogen bond is disrupted, so a hydrogen atom is also released. The acetic acid thus formed combines with coenzyme A, a derivative of pantothenic acid (a B vitamin)producing the compound acetyl coenzyme A (acetyl CoA). Naturally, the whole process happens with the components or reactants immersed in water.

Acetyl CoA then enters the citric acid cycle, which consists of a cyclical series of eight separate biochemical reactions, all of them immersed in water; that are directed by the enzymes of the mitochondrial matrix. In spite to be described as a cyclical series of biochemical reactions, the molecules themselves are not physically moved around in a cycle.

This cycle is alternatively known as the Krebs cycle or the tricarboxylic acid cycle, because citric acid contains free carboxylic acid groups. As two carbons are removed from the six-carbon citric acid molecule, converting it back into the four four-carbon oxalo-acetic acid, which is now available at the top of the cycle to pick up another acetyl CoA for another revolution through the cycle.

The released carbon atoms, which were originally present in the acetyl CoA that entered the cycle, are converted into two molecules of CO_2. This CO_2 as well the CO_2 produced during the formation of acetic acid from pyruvic acid passes out of the mitochondrial matrix and subsequently out of the cell to enter the blood. In turn, the blood carries it to the lungs, where is finally eliminated into the atmosphere through the process of breathing. The oxygen used to make CO_2 from this released carbon atoms is derived from the molecules that were involved in the reactions, not from free molecular oxygen supplied by breathing.

Pyruvic acid is the end product of anaerobic glycolysis and is converted to lactic acid when it cannot be further processed by the oxidative phosphorylation pathway. Pyruvic acid must pass through membranes composed mainly of lipids and proteins plus a small amount of carbohydrates. The most abundant membrane lipids are phospholipids, with lesser amount of cholesterol. An estimated billion phospholipids' molecules are present in the plasma membrane average. Phospholipids have a polar or electrically charged head containing a negatively charged phosphate group and two non/polar –electrically neutral- fatty acid tails. The polar end is hydrophilic because it can interact with water molecules, which are polar also; the non-polar end is hydrophobic and will not mix with water. The hydrophobic tails bury themselves in the center away from the water.

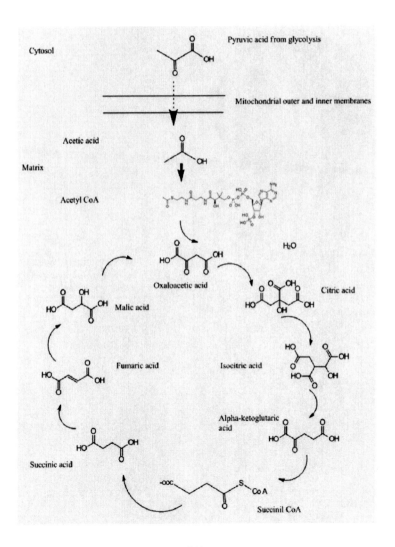

The lipid bilayer is not a rigid structure but instead is fluid in nature, whit a consistency more liquid cooking oil than like solid shortening. The phospholipids, which are not held together by strong chemical bonds, are able to twirl around rapidly as well as move about within their own half of the layer, much like skaters along a crowded skating rink.

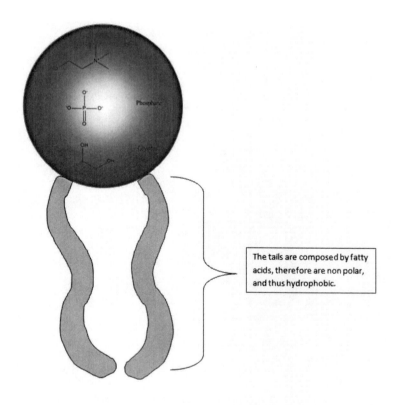

The tails are composed by fatty acids, therefore are non polar, and thus hydrophobic.

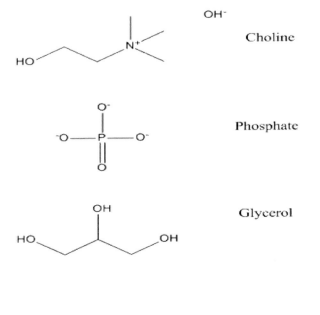

Choline

Phosphate

Glycerol

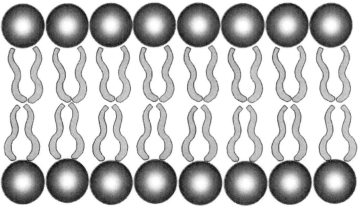

Diagram of lipid bilayer, the main component of the cell membrane.

These are the components of the head, which are polar and hydrophobic:

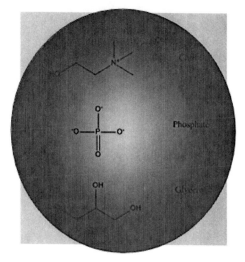

And this is the formula of phosphate group that is highly regulated by diverse reasons and by several mechanisms, for instance ATP and GTP.

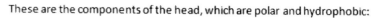

Notice the remarkable presence of phosphate´s groups within the cell membrane as **phospho**lipids.

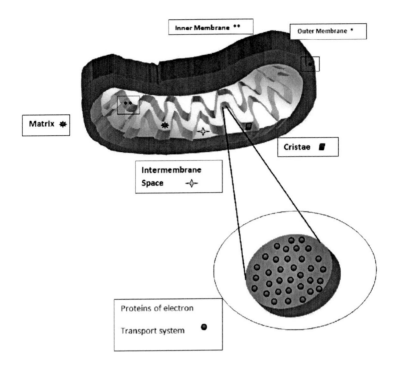

Mitochondrial Matrix

In recent years it has become apparent that biological systems contain high concentrations of macromolecules and thus differ markedly from the dilute in vitro solutions typically used in experimental studies. No biological system better exemplifies this point than the mitochondrial matrix, which is the site of many important oxidative reactions, including those involved in the tricarboxylic acid cycle and fatty acid oxidation pathway. It has been suggested that the concentration of protein in the matrix is so high that it may approach the close-packing limit, thereby creating an unusual "aqueous" milieu in which the motion of metabolic substrates and protein is unusually hindered thus requiring more energy.

Relatively little is known about the mitochondrial matrix, and elucidating the relationship between metabolism and matrix properties thus remains a key problem in biochemistry. The mitochondrial configurations −condensed and orthodox—are dependent on the osmolarity of the external medium and on respiratory state, whether or not the mitochondria are active in electron transport and ATP synthesis.

Unfortunately, it has been difficult to proceed beyond the level of ultra-structure to study matrix properties at molecular level. One of the major impediments to such research has been the fact that the matrix is surrounded by the outer and inner mitochondrial membranes, the latter of which is highly impermeable. Thus, it has been difficult to introduce chemical probes into the matrix of intact functional mitochondria, and generally one has been left with the alternative of disrupting the mitochondria partially or completely to study matrix properties that modify in an unknown manner both partially or completely the matrix properties.

Thereby attempts to measure an effective viscosity for the matrix and therefore to identify the effects that high matrix protein concentrations have on matrix structure and function, taking into account that is an environment in which free space is limited. At highest osmolarity (300-400 mOsm) the mitochondria were in the condensed configuration and the anisotropy was large. As the osmolarity was lowered the mitochondria became more orthodox in appearance and the anisotropy fell measurably, indicating that diffusion became progressively more rapid as the matrix protein concentration decreased. Thereby in order to build-up a highly concentrated matrix protein, both hydrogen and water diffusion must be lowered, for instance with two membranes and one space between them.

Below about 120 mOsm or less, structural alteration occurred in which the outer membrane was ruptured and the matrix compartment was highly swollen. At this point the anisotropy began to fall markedly. When the osmolarity was decreased to about 40 mOsm or less, the inner membrane unfolded completely into spherical configuration, the matrix density decreased further and between 25 mOsm and 2 mOsm the value of the anisotropy approached zero[17]. The physiological osmolarity is 300 mOsm. Milliosmole is 0^{-3} part of osmole, a unit of osmotic pressure equivalent to the amount of solute that dissociates in solution to form one mole (Avogadro´s number) of particles (molecules and ions). In other words: osmole is the quantity of a substance in solution in the form of molecules, ions, or both (usually expressed in grams) that has the same osmotic pressure as one mole of an ideal non-electrolyte, also spelled osmol.

During electron transport driven y the oxidation of an appropriate respiratory substrate such as succinate, diffusion in the matrix compartment was moderately hindered. In contrasts, when ATP synthesis was initiated by adding ADP, the mitochondria adopted a more condensed configuration, matrix density increased, and diffusion of fluorescent probes as carboxyfluorescein was correspondently more hindered. Finally, when electron transport was inhibited by adding antimycin A, the mitochondria adopted a highly condensed configuration, and carboxyfluorescein diffusion was hindered further.

The mitochondrial matrix is around 20 times more viscous than cytoplasm. This viscosity data suggest the time scale associated with the orientation of a substrate with respect to an enzyme will be lengthened considerably in the intact matrix. It is usually observed that translation is more hindered in concentrated solutions than is rotation. Thus, the data suggest that translation and collisional interactions in the mitochondrial matrix will also be significantly impeded, especially those as acetyl-CoA, $FADH_2$ and NADH, which exceed carboxyfluorescein in size. And since all metabolites must associate with and dissociate from enzymes, processes which require energy; it is anticipated that the hindrance of motion may be even more pronounced for some matrix metabolites.

Matrix viscosity is volume and/or configuration dependent. It appears that a number of matrix enzymes turn over more rapidly as the volume of the matrix increases in response to hormonal stimuli. Metabolite diffusion is involved in the mechanism of action of these enzymes, and then such enhanced turnover can reflect more rapid diffusion of metabolites that is brought about by a reduction in viscosity as matrix volume increases.

If the diffusion of small molecules were in more or less degree inhibited in the crowded matrix environment, one might infer that channeling of substrates from one enzyme to another in sequence occurs. However, channeling; although it may occur, seems it's as not essential consequence of a crowded matrix environment.

It has been suggested that the matrix is a dense-packed protein solution in which the average diameter of protein-free aqueous pores is only 2 nm, (Hydrogen atom measures 1 nm); and in which water has changes in its properties. It appears that matrix water is 3-4 times less mobile than bulk water and can even exist in two phases: one bound to protein or membrane and highly viscous and the other free and with a viscosity equal to that dilute buffer.

In general sense, data support the concept of a dense-packed matrix in which the space available for diffuse motion is limited. This concept is consistent with our observation that diffusion is severely hindered in the aqueous phase of the condensed matrix and that diffusion accelerates as the matrix becomes orthodox and its water content and pore size increases, processes that require energy, indeed. The matrix protein in a condensed mitochondrion is so tightly packed that further structural change is unlikely to occur.

The high viscosities and slow diffusion detected by some researchers can arise either as a weighted contribution from two distinct phases, such as bound and free water with relatively higher and lower viscosities, or from a single aqueous phase that actually has the measured viscosity. Whatever the exact structure of the matrix may be, it is now becoming clear that the molecular dynamics of the growing spheres of diatomic hydrogen and the flow of high energy electrons; besides the structure and function in the mitochondrial matrix are very profoundly influenced and modulated in several ways by variations in matrix protein concentration.

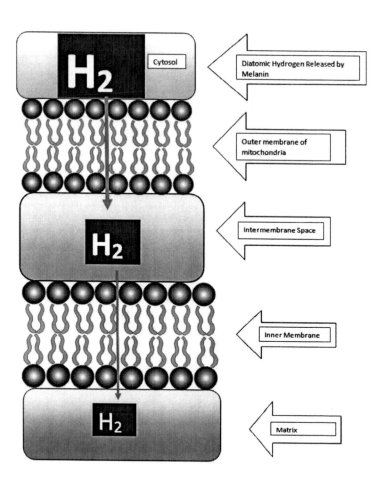

The inner mitochondrial membrane is highly impermeable.

Presently is believed that the source of energy for the body is the chemical energy stored in the carbon bonds of ingested foods.

Thereby the cell burns glucose combining it with oxygen, and the "smoke" is the CO_2.

Incongruously, text-books also said: Body cells are not equipped to use this energy directly, therefore they must extract energy from food nutrients (burning them) and the energy released through the combination of Oxygen with glucose, it's stored in the high-energy phosphate bonds of the adenosine Triphosphate.

**Adenosin
triphosphate
ATP**

Chemical Formula: $C_{10}H_{16}N_5O_{13}P_3$
Exact Mass: 507.00
Molecular Weight: 507.18
m/z: 507.00 (100.0%), 508.00 (11.5%), 509.00 (3.5%), 507.99 (1.8%)
Elemental Analysis: C, 23.68; H, 3.18; N, 13.81; O, 41.01; P, 18.32

However, thermodynamic laws said that energy cannot be stored.

Theoretically the process is as follows:

Supposedly energy is released symmetrically in all directions when ATP is degraded to ADP. But ATP is very unstable in water and the reaction occurs immersed in aqueous medium, therefore "energy" cannot be stored in that way. By other side, in reality when ATP is broken down to ADP energy is <u>absorbed</u> as follows:

Notice: Energy is <u>absorbed</u> when ATP is downgraded to ADP. Probably the energetic level of molecules nearby to ATP are lowered and this can be the explanation for their apparent impeling function due to the changes in energy of any molecule also induces a response in one or another way. It is much more easy to lower the energetic level of any estructure than the opposed: the increasing of the energetic level. For instance for each 10 receptors in the cell membrane that increase the energetic level, exist 90 receptors in the cell membrane that decrease the processes involved in it. It is hardest the increasing than the lessening of the energetic level.

It is probably that one of the main functions of ATP is the regulation of the phosphate groups, because no substance can be floating freely in the cytoplasm, except small molecules as diatomic hydrogen and H_2O. ATP is a molecule that is able to give and take phosphate groups so appears to function as metabolic regulator.

The main building block of the body: glucose; is in higher concentration in the blood than in the tissues. Fresh supplies of this nutrient are regularly added to the blood by eating and from glucose reserve (biomass) stores in the body. Simultaneously, the cells metabolize glucose almost as rapidly as it enters the cells from the blood. As a result, a continuous gradient exist for net diffusion of glucose into the cells. However, glucose cannot cross cell membrane on its own. Glucose, being polar, it is not lipid soluble, and it is too large to fit through a channel. Without the glucose carrier molecules to facilitate membrane transport of glucose, the cells would be deprived of glucose, their preferred source of biomass.

Thereby human photosynthesis is a relatively simple mechanism by which all cells might obtain and use energy to drive active and passive transport (keeping the shape of the molecules involved in it) of all solutes without necessity of conserve it in the form of adenosinetriphosphate (ATP).

Aneuploidy and Cancer

Cancer cells often show Aneuploidy, a state in which cells have an abnormal number of chromosomes. Aneuploidy led to other genetic abnormalities in the cells similar to the sort of "genomic instability" that is a hallmark of cancer cells[18]. Whole-chromosome Aneuploidy or a karyotype that is not a multiple of the haploid complements is found to greater than 90 % of human tumors and may contribute to cancer development. Aneuploidy increases genomic instability and a consequence of the later is the worsening of the Aneuploidy itself.

When the body makes a mistake, for instance a wrongly products of some apparent metabolic pathway means that many things must happened erroneously, a very rare fact in processes finely tuning along four billion years of evolution by Nature.

In my opinion human body is almost perfect, and it is not programmed to make mistakes or goes under disease, instead, under given circumstances, the body makes its best effort; thereby the illnesses are not body´s miscalculations, instead are probably the best options in a body that had lost the energetic balance.

CHAPTER 17

The Human Photosynthesis and Storage diseases

The storage diseases are a group of metabolic disorders with certain characteristics in common. One of the most remarkable is in regards to long carbon chains that are a Carbon property, which makes the longest carbon chain in comparison with other elements of the periodic table. By other hands all storage diseases originate from an abnormal accumulation of substances, by example: in lysosome.

Depending of the number of Carbon atoms, the compounds are called: Lipids, mucopolysaccharides or glycoprotein.

Lipids are fat-like substances that are important parts of the membranes found within and between each cell ad in myelin sheath that coats and protects the nerves. Lipids include oils, fatty acids, waxes, steroids (such as cholesterol and estrogen) and other related compounds.

Gaucher disease is the most common of the lipid storage diseases. It appears to be caused by a deficiency of the enzyme glucocerebrosidase. Fatty material can collect in the spleen, liver, kidneys, lungs, brain and bone marrow. Symptoms may include enlarged spleen and liver, liver malfunction, skeletal disorder and bone lesions that cause pain and fractures, severe neurologic complications, swelling of lymph nodes and (occasionally) adjacent joints, distended abdomen, brownish tint of skin, anemia, low blood platelets, and yellow spots in the eyes. There is an increase in the susceptibility to infection. The disease affects males and females equally.

Niemann-Pick disease is actually a group of autosomic recessive disorders caused by an accumulation of fat and cholesterol in cells of the liver, spleen, bone marrow, lungs, and in some patients, brain. Neurological complications may include ataxia, eye paralysis, brain degeneration, learning problems, spasticity, feeding and swallowing difficulties, slurred speech, loss of muscle tone, hypersensitivity to touch, and some corneal clouding. A characteristic cherry-red halo develops around the center of the retina in 50 % of patients.

Fabry disease, also known as alpha-galactosidase-A deficiency, causes a buildup of fatty material in the autonomic nervous system, eyes, kidneys, and cardiovascular system. Fabry disease seems as the only X-linked lipid storage disease.

Farber´s disease or Farber´s lipogranulomatosis, describes a group of rare autosomic recessive disorders that cause an accumulation of fatty material in the joints, tissues, and the central nervous system. The disorder affects both males and females.

The gangliosidoses are comprised of two distinct groups of genetic diseases. The GM1 gangliosidoses are caused by a deficiency of the enzyme beta-galactosidase, resulting in abnormal storage of acid lipid materials particularly in the nerve cells in the central ad peripheral nervous system. The GM2 gangliosidoses also cause the body to store excess acidic fatty materials in tissues and cells, most notably in nerve cells. These disorders result from a deficiency of the enzyme beta-hexosaminidase. The GM2 disorders include: Tay-Sachs disease, Sandhoff disease; Krabbé disease, also known as globoid cell leukodystrophy and galactosylceramide lipidosis; Metachromatic leukodystrophy or MLD, Wolman´s disease, also known as acid lipase deficiency.

Fatty material can collect in the spleen, liver, kidneys, lungs, brain, and bone marrow. Symptoms may include enlarged spleen and liver, liver malfunction, skeletal disorders and bone lesions that may cause pain and fractures, severe neurologic complications, swelling of lymph nodes and (occasionally) adjacent joints, distended abdomen, a brownish tint to the skin, anemia, low blood platelets, and yellow spots in the eyes. Symptoms may begin early in life or in adulthood and include enlarged liver and grossly enlarged spleen, which can rupture and cause additional complications. Skeletal weakness and bone disease may be extensive. The brain is not affected, but there may be lung and, rarely, kidney impairment. Symptoms include an enlarged liver and spleen, abnormal eye movement, extensive and progressive brain damage, spasticity, seizures, limb rigidity, and a poor ability to suck and swallow. Major symptoms include an enlarged spleen and/or liver, seizures, poor coordination, skeletal irregularities, eye movement disorders, blood disorders including anemia, and respiratory problems. The accumulation of fat and cholesterol usually are in cells of the liver, spleen, bone marrow, lungs, and, in some patients, brain. Neurological complications may include ataxia, eye paralysis, brain degeneration, learning problems, spasticity, feeding and swallowing difficulties, slurred speech, loss of muscle tone, hypersensitivity to touch, and some corneal clouding, develop an enlarged liver and spleen, swollen lymph nodes, nodes under the skin (xanthemas), and profound brain damage by 6 months of age. The spleen may enlarge to as much as 10 times its normal size and can rupture. These children become progressively weaker, lose motor function, may become anemic, and are susceptible to recurring infection. Most patients also develop ataxia, peripheral neuropathy, and pulmonary difficulties that progress with age, but the brain is generally not affected. Neurological signs include burning pain in the arms and legs, which worsens in hot weather or following exercise, and the buildup of excess material in the clear layers of the cornea (resulting in clouding but no change in vision). Fatty storage in blood vessel walls may impair circulation, putting the patient at risk for stroke or heart attack. Other manifestations include heart enlargement, progressive kidney impairment leading to renal failure, gastrointestinal difficulties, decreased sweating, and fever. Angiokeratomas (small, non-cancerous, reddish-purple elevated spots on the skin) may develop on the lower part of the trunk of the body and become more numerous with age. These symptoms may include moderately impaired mental ability and problems with swallowing. The liver, heart, and kidneys may also be affected. Other symptoms may include vomiting, arthritis, swollen lymph nodes, swollen joints, joint contractures (chronic shortening of muscles or tendons around joints), hoarseness, and xanthemas which thicken around joints as the disease progresses. Patients with breathing difficulty may require insertion of a breathing tube.

The clinical picture may include neurodegeneration, seizures, liver and spleen enlargement, coarsening of facial features, skeletal irregularities, joint stiffness, distended abdomen, muscle weakness, exaggerated startle response, and problems with gait. About half of affected patients develop cherry-red spots in the eye. Children may be deaf and blind by age 1 and often die by age 3 from cardiac complications and pneumonia. Symptoms begin by 6 months of age and include progressive loss of mental ability, dementia, decreased eye contact, increased startle reflex to noise, progressive loss of hearing leading to deafness, difficulty in swallowing, blindness, cherry-red spots in the retinas, and some paralysis. Seizures may begin in the child's second year. Children may eventually need a feeding tube and they often die by age 4 from recurring infection. Neurological signs may include progressive deterioration of the central nervous system, motor weakness, early blindness, marked startle response to sound, spasticity, myoclonus (shock-like contractions of a muscle), seizures, macrocephaly (an abnormally enlarged head), and cherry-red spots in the eye. Other symptoms may include frequent respiratory infections, murmurs of the heart, doll-like facial features, and an enlarged liver and spleen. Other symptoms include muscle weakness, hypertonic (reduced ability of a muscle to stretch), myoclonus seizures (sudden, shock-like contractions of the limbs), spasticity, irritability, unexplained fever, deafness, optic atrophy and blindness, paralysis, and difficulty when swallowing. Prolonged weight loss may also occur. Symptoms include impaired school performance, mental deterioration, ataxia, seizures, and dementia. Symptoms are progressive with death occurring 10 to 20 years following onset. In the *adult* form, symptoms begin after age 16 and may include impaired concentration, depression, psychiatric disturbances, ataxia, seizures, tremor, and dementia. Develop progressive mental deterioration, enlarged liver and grossly enlarged spleen, distended abdomen, gastrointestinal problems including steatorrhea (excessive amounts of fats in the stools), jaundice, anemia, vomiting, and calcium deposits in the adrenal glands, causing them to harden.

In general terms, we could say that storage diseases have common signs and symptoms.

	Gaucher	Niemann-Pick	Fabry	Farber	GM1	GM2
Spleen	•	•	•	•	•	•
Liver	•	•	•	•	•	•
Kidney	•	•	•	•	•	•
Brain	•	•	•	•	•	•
Bone Marrow	•	•	•	•	•	•
Bone	•	•	•	•	•	•
Blood	•	•	•	•	•	•
Eye	•	•	•	•	•	•
Swallowing	•	•	•	•	•	•
Heart	•	•	•	•	•	•

The failure in these diseases is generalized, therefore we must think in energy first.

In regards to the arrangement of carbon chains, characteristically long chains of carbon are the mains constituent of the fatty deposits, in example we have:

$$C_{67}H_{121}N_3O_{26}$$

Ganglioside GM2; Tay-Sachs ganglioside; β-D-GalNAc-(1→4)-[α-Neu5Ac-(2→3)]-β-D-Gal-(1→4)-β-D-Glc-(1↔1)-*N*-octadecanoylsphingosine.

Other good example is ceramide with the following formula:

In reality the growing of the carbon chains is relatively easy because is until certain points a natural trend of the carbon itself, and the chemical mechanisms of our body tend to drive this trend in order to make chains with the adequate extension, branching; and combination with other elements as oxygen, nitrogen, hydrogen; etc. Therefore is easiest for the cell, tissue, organ, or body; drive the growing of the molecule than promotes the growing itself. Finally could be the combination of the several available biochemical mechanisms that gradually drives the conformation of the final molecule. In any case the very first requirement is the same: energy.

Glycogen storage disease (GSD, also glycogenosis and dextrinosis) is a good example in which an energy failure is a very plausible explanation. GSD is the result of defects in the processing of glycogen synthesis or breakdown within muscles, liver and other cell types. Either glycogen synthesis or breakdowns are biochemical processes that have a very first requirement in common: energy. GSD has two classes of cause: genetic and acquired. In livestock, acquired GSD is causes by intoxication with the alkaloid castanospermine.

Our research finally broke the logjam about the main source of energy of eukaryotic cell, however, by other side open a new field of research and therapeutic possibilities. A body with four billion years of evolution is highly complex, and the best of all: is a living thing, therefore the order, sequence, temporality, spatial order and other known and unknown characteristics of the myriad of biochemical reactions comprised in the metabolism of our organism are so far of our abstraction capacity, however, the very first reaction in common finally was unraveled; from now in ahead we could think in different options in regards to storage diseases, since pathogenesis and treatment point of view.

Summarizing: The very first requirement in order that our body drives adequately carbon chains length: energy.

CHAPTER 18

The Eye and Oxygen

In the early 1950s there was a vague feeling that the key to solving the mystery of Retrolental Fibroplasia, might lie in improved understanding of the unusual way in which the retinal layer of the eye receives its blood supply during fetal development. Ida Mann, in 1928, described the embryonic development of the human eye and developed the concept that the retinal vessels originate by budding from the base of the fetal intraocular blood vessel of the eye: the hyaloids artery.

At the end of the 1940s, Michaelson injected India ink into the arterial system to fill and blacken the smallest vessels of the fetal eye, removing the globe for dissection, and then teasing out the retinal layer to make a flat preparation which he mounted on a glass slide.

In accordance to the view established by these observations, the human retina has no blood supply of its own until the third month of fetal life.

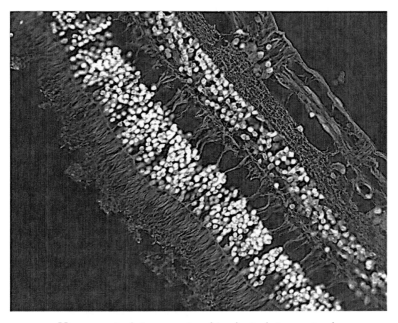

Human retinal tissue section, histological view stained
with H & E, 100 X microphotography.

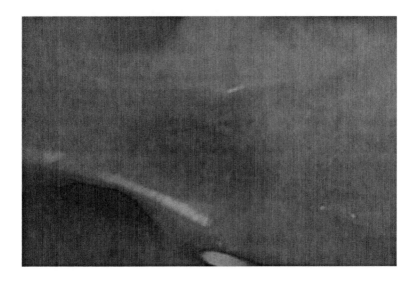

Mongolian spot

This was interpreted to indicate that retinal tissue receives adequate amounts of oxygen and nutrient from the vessels of the choroid layer of the eye which lies just beneath the retina. But in accordance with our unraveling about the intrinsic property of melanin to split and reform the water molecule, then the adequate amounts of energy to the growing retinal tissue comes from melanin placed mainly in the choroidal layer of the eye. Michaelson suggested that as the retina develops and becomes thicker at the third to fourth month of gestation, the nutritional needs can no longer be met by the nearby choroid. However, every single eukaryotic cell possesses its own photosynthesis system and retinal tissue is not an exception. In my opinion the gradual development of a circulatory system reflects mainly the imperiously necessity to drive the expelling of CO_2.

The circulatory system form in response to the growing CO_2 transport needs of the mature tissues.

The classical experiment of Norman Ashton, a pathologist of the Institute of Ophthalmology, University of London, where Ashton showed that after few days in continuous high oxygen (60-70 percent concentration), the outgrowing retinal vessels was completely withered; and that can be explained in terms of an imbalance in the equation of the human photosynthesis as follows:

$$2H_2O \leftrightarrow 2H_2 + O_2 + 4e^-$$

As any chemical reaction, reactant concentrations is a very important factor, therefore when the level of diatomic oxygen in the surroundings is more than 20-22 percent, then the equation will trend to the left, i.e. water molecule reformation is abnormally increased and thereby diatomic hydrogen is consumed, in other words energy levels are lowered and the highly requirements of a growing tissue are impaired. The imbalance in the equation can schematized as follows:

$$2H_2O \leftrightarrow 2H_2 + O_2 + 4e^-$$

Hydrogen consumption is highest than the amount of produced hydrogen by mean of water dissociation. In spite that is a way to protect tissues versus the excessive levels of the toxic oxygen, due to the low levels of diatomic hydrogen, the energy carrier by excellence; then the available energy is not enough to impel the myriad of chemical reactions that happen constant and simultaneously in every single cell of the body, thereby the physiological, normal and even abnormal processes are modify in more or less degree in an effort of the body trying to keep or maintain the fine tuning dynamic balance between energy and biomass developed along four billion years of evolution in the best way; but sooner or later some sign or symptom of enough magnitude to be detected by our senses will happen.

The Norman Ashton team conclusion about that constriction of immature blood vessels was the primary effect of constant high oxygen levels is remarkable, but undoubtedly explanation is much more complex. When oxygen levels are maintained high for a long enough periods, obliteration of the entire developing blood vessel network of the retina took place. Regrowth of blood vessels after return to air oxygen levels took place in a wild disorganized fashion with budding of new capillaries into the vitreous portion of the eye in front of the retina, a consistent finding with the early blood vessels changes in premature retinopathy. An organism with greater melanin content is able to support in better way and for more extended periods of time the imbalance in the equation, which is congruous with Phelps observation that Premature retinopathy is less severe on black people (Phelps 1997).

Ironically and paradoxically, it was exposure to high oxygen concentration, with resultant obliteration of developing retinal blood vessels, which appeared to cause the ultimate oxygen deficiency in the deep retinal layer of the eye. These results seemed bizarre: breathing high concentrations of oxygen appeared to render one tissue of the body oxygen-impoverished. And by other, the photoreceptor layer, the most external layer of retinal tissue properly that is adjoin to pigmented cells, has not blood vessels at all; in spite that requires ten times more energy than cerebral cortex, six times more energy than cardiac muscle and three times more than kidney cortex. In their experiments, the Norman Ashton team, without noticing produces experimentally imbalance of the equation in both senses, and with low levels of oxygen usually was not sufficient to give rise to proliferation of retinal blood vessels. The imbalance in the equation produced by low levels of oxygen can be schematized as follows:

$$2H_2O \leftrightarrow 2H_2 + o_2 + 4e^-$$

And the explanation is that reaction will tend to the left, so the process of water dissociation will be favored, increasing the presence of diatomic hydrogen, but the high energy electron levels will be lowered, which has consequences, because both kinds of energy are necessary to impel biochemical reactions that are different among them and even opposite under certain circumstances. For instance when the reaction is affected by high levels of oxygen, for instance, breathing high concentrations of oxygen; is as follows:

$$2H_2O \leftrightarrow 2H_2 + O_2 + 4e^-$$

The characteristic clinical and histological finding in this case is vasoconstriction in the peripheral retina, which appeared to cause severe oxygen deficiency in the deep retinal layer of the eye, but not in photoreceptors layer because this part of the tissue normally has not blood vessels at all.

By other side, when the reaction is misbalanced by low levels of oxygen, as in the following equation:

$$2H_2O \leftrightarrow 2H_2 + o_2 + 4e^-$$

The reaction of the retinal blood vessels to higher levels of diatomic hydrogen, lower levels of water, diatomic oxygen and high energy electrons is not to the proliferation, appearing that this adjusted metabolic condition is better supported by the retinal tissue, which is congruous with our discovery of the intrinsic property of melanin to split and reform the water molecule due to the cells can produce diatomic oxygen at their own by mean of the water molecule dissociation. In my opinion the main function of the diatomic oxygen collected by lungs, it is not as a simple burning co-fuel to be used with glucose, instead oxygen is a part of the equation that allows the body an adequate drive and expelling of CO_2 mainly, but not unique, through the effect of the diatomic oxygen a very stable molecule; over hemoglobin, recall that hypoxia worsening hypercapnia. Low levels of oxygen and therefore the low levels of high energy electrons are better tolerated by the cell than low levels of molecular hydrogen induced by higher levels of diatomic oxygen.

In 1953, Arnall Patz and his co-workers conducted extensive studies of high oxygen exposure in several species of newborn animals, beginning with opossum and rats and later with mice, kittens, and puppies. Their findings exactly parallel the early stages of the human disease. The difference in subsequent course is dependent upon the development of retinal detachment which did not occur in the kitten and other experimental animals. In general terms the experimental animals did not become blind as result of high levels of oxygen, oppose to evolution in the human newborn, in which the final result is blindness.

It is understandable that in previous works to my discovery about human photosynthesis, researchers did not take in account that amount of melanin as a risk determinant factor. In my experience a greater amount of melanin is a protection factor versus retrolental fibroplasia or premature retinopathy, because with higher amounts of melanin, the body is able to regain easiest the balance of the central equation that constitutes the origin, support and that governing the body as a whole.

In the previous research about retrolental fibroplasia or premature retinopathy, Patz et al, made the same experimental protocol in different species, but they didn´t take into account differences in the amount of melanin among them. For instance, in the work of Gunn, Easdown, Outerbridge and Aranda[19] about possible determinant factors that may increase the risk of the occurrence of retrolental fibroplasia (RFL) were analized in 80 infants born in 1975 and 1976 with birth weights between 501 and 1500 g, and who survived. Active and/or cicatricial RFL occurred in 27 (33.8 %) infants and the factors significantly associate with RLF were: gestational age; apnea requiring bag and mask resuscitation with oxygen, septicemia, degree of illness, blood transfusion; and mechanical ventilation. The amount of melanin of the infants born studied is not reported in the Gunn´s study, due to melanin is not considered an important factor, because it is only a "sunscreen". Recall that in certain times of the history of the human kind, the books, all of them; said that the earth was flat.

Returning to Premature retinopathy or RFL, when the skin color is taken into account, then difference emerge. In the study of Saunders, Phelps et al[20], which objective was determine and compare the incidence of severe, vision-threatening retinopathy of prematurity (ROP) in black and white low-birth-weight infants; and their conclusions were severe, vision threatening ROP occurs with greater frequency in low-birth-weight white infants (7.4 %)

than in low-birth-weight black infants (3.2 %) who are seemingly at equivalent risk. Saunders and Phelps concluded that reason for this disparity is unknown. Recall that in 1997 the intrinsic property of melanin to split and reform the water molecule was unknown.

It is expected that both diatomic hydrogen with its precious cargo of energy and high energy electrons, released both by the human photosynthesis system, are able to impel different parts of the same biochemical reaction or even in different manner, because the energy carried by diatomic hydrogen has several differences with the energy that the flow of high energy electrons possesses. It is perfectly possible that the same molecule or substrate can react chemically different depending of the predominant nearest type of energy.

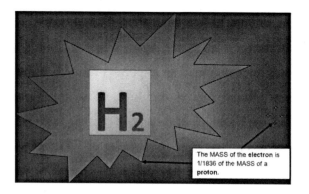

The MASS of the electron is 1/1836 of the MASS of a proton.

The differences in energy are a very complicated question with no simple answer. In energy units (using $E = mc^2$), the masses are: Proton: 938.272 MeV, neutron: 939.566 MeV, mass difference = 1.293 MeV, electron: 0.511 MeV.

In regards to the eye, only the incompletely developed retina is susceptible to this effect of hyperoxia. Once blood vessel development is complete, supplemental oxygen causes a decrease in the caliber of the mature retinal vessels, something similar happens with the blood vessels of the optic nerve; but pronounced vasoconstriction, obliteration and finally angiogenesis are events´ chain that has resemblance with both ROP and proliferative diabetic retinopathy.

The effects of melanin over body´s balance are highly complex and involve the whole organism, and not only the eye. For instance, babies of African mothers are born at a gestation five days shorter than their European counterparts with a substantially less incidence of respiratory distress syndrome in the preterm African babies, compared with their European counterparts[21]. Recall that sun light is 30% more brilliant in Africa than in Europe.

CHAPTER 19

Human Photosynthesis and the other areas of Knowledge

Melanin's extraordinary property to unfold the water molecule, unknown until today is a knowledge that must shape a whole new era in all fields of human science.

Let's see a few ones:

ON ENERGY GENERATION

Today, the dream of producing electricity not burning oil, without generating greenhouse's gases, without the dangers of nuclear centrals and its wastes. Energy's alternative generation is already occurring. In biology, living organism get energy through several processes, but getting energy from electromagnetic radiations (light's visible spectrum) hasn't been considered on living organisms, be them uni or pluricellulars, and some algae, fungus and cianobacterium.

In the animal kingdom, we have only studied exhaustively the usual metabolic pathways through which, with the degradation of different molecules (lipids, proteins, cations and anion glycons) that enter the organism with food, different kinds of energy are generated, TAP being the universal exchange currency. But until now, the possibility of living organisms being able to capture electromagnetic radiations on a meaningful way and/or to transform it in useful energy for every process required to life and to have an optimal development of the senses that allow us to interact with our environment. This property from dark or black compounds, generally called melanins (eumelanins and to a lesser degree, pheomelanins) has been unnoticed, but if we apply this new knowledge in the study of cells, tissues, organs and systems both its physiology and pathology, a whole new horizon opens before us, one that allows to explain and/or to understand more deeply some of the many mysteries of melanin's tenacious and/or persistent existence in the immense majority of living beings.

Melanin are usually described as substances whose function inside the organism if focused to or comes from the protection against ultraviolet radiations in the skin, but since now we accept this compound is not as protective or impregnable against these radiations as thought, the pigment's presence inside organs and inner structures remained an enigma, and even worse, it wasn't supposed to have any importance at all.

However, melanin's ability to absorb the whole visible spectrum, plus its ability to use this energy to break the water molecule, generating oxygen and hydrogen with it, let us begin to understand why this pigment is so present on places exposed to the sun. However, the purpose is not to absorb radiations only to protect, but to use it productively, generating energy with the hydrogen and the oxygen, something really important to aerobic organisms.

This greater oxygen availability can be proved noticing that, the darker is the skin; the lower is the blood's hemoglobin concentration. Even more, in the scientific literature can be found this, even 20 gram of difference between whites and blacks.

Yet another data supporting melanin's photolytic properties is the fact that lungs are smaller the darker the skin gets. If we take notice of melanin's photolytic function we can complete and begin to understand some mysteries, like aqueous humor dynamics inside the eye, making it (aqueous humor synthesis, secretion, and excretion) less difficult to understand. If we accept melanin do intervene in a meaningful way in aqueous humor secretion, transport, and excretion, we begin to build a more coherent picture about the reasons why intraocular pressure is higher in the mornings, given its secretion descends 45% during night, understandable since, without light, melanin that cover the ciliary body (a tissue produced by the humor), has less energetic efficiency to secrete and to excrete, because the lack of light at night results in a 45% descend on aqueous humor production, as well as for excretion, since the energy melanin allows to take from light, also intervenes in excretion processes (aqueous humor output), because during night, despite the aqueous humor production descending almost by half, intraocular pressure slowly grows until dawn, when is usually the highest pressure during all day.

Curiously, as day goes by and the sun lightens up the eye's pigmented tissues for hours, then intraocular pressure descends constantly until dusk, then intraocular pressure is the days lowest, despite the aqueous humor production grows in 45% during day, that is, during light hours. However, this grow in aqueous humor production possibly following a bigger light amount and, hence, a bigger energy amount is compensated, by a greater efficiency on the mechanisms on charge of the evacuation from inside the eye to the exterior, thanks to a higher energy availability in hydrogen and oxygen shape, which is probably included in metabolic processes by NADH and NADPH, as well as by oxygen used for multiple purposes in which it intervenes. As an example, we have electrons expelled from the metabolic pathway called oxidant phosphorylation gathered, this comes from several reduced carbon forms catabolism, which we ingest and that represent our main energy source. We exhale this carbon as carbon dioxide at the end of metabolic processes, carbon most oxidized form. This final compound (CO_2), exhaled or expelled by organisms, is the completely opposite to what happens with plants, for they absorb CO_2 and water, and, with sunlight, they form carbon hydrates (glucose) and we absorb glucose and produce water and CO_2.

This radical difference is but one of the particularities that made us think animal beings were far away plants, and, hence, photosynthesis only applied to them (plants) and never on us (animals), because we don't have chlorophyll and being completely different to each other. However, our discovery of melanin's photolytic properties when in water is going to revolutionize these concepts, causing a big impact on cellular biology as a whole.

Yet another example on melanin's photolytic activity importance appears when we study light's transduction process, which occurs inside mammals' retina photoreceptor cells and partly by chance? In one of the organism most pigmented cells, the retina's pigmented epithelium, in textbooks analyzing this phenomenon (transduction), name a step in the biochemical process, unknown in origin (according to books) is the pigmented epithelium cells descend on intracellular pH when lighted up. These descend opens chlorine channels, which begins the processes needed to perceive light.

Melanin has all the characteristics to be the mysterious "unknown" substance (Adler, 1998) absorbing light and rising hydrogen ions availability in cytosol, not knowing its origins, because it doesn't seems to be some other substance degrading it and releasing this tiny element. However, if we take notice that melanin only acts as a catalyze and uses water as primary resource, which is abundant in the surroundings, then the circle seems complete, because, besides being a light energized process, it doesn't has an ATP expense, which is also congruent with the described metabolic processes.

This photolytic properties of melanin, explains why the blood's oxygen concentration circulating in the eye's most vascular layer, called choroids or uvea, is the entrance (choroidal circulation arterial side) of 97%, and 94% at the exit (choroidal circulation venous side) something unique in the whole body, since the carbon dioxide concentration is similar to the rest of the body, 40% (in the venous side). According to literature, carbon dioxide stimulates water's photolytic process, but it's unknown if this occurs through a melanocytes' stimulation or by melanin itself. We refer the reader to the patent solicitude GT/a/2005/00006 for detailed description of the way in which melanin capture electromagnetic radiations and generate hydrogen and oxygen from water, as well as the opposite action, that is, the reunion of oxygen and hydrogen, producing water and generating electricity.

Photosynthetic microbes, as green algae and cianobacteria, extract hydrogen from water as part of their metabolic activities using luminous energy as their main source. However, as oxygen gets produced at the same time as hydrogen, the sensibility to oxygen of these enzymatic systems may be what limits this microorganisms hatching. Besides, hydrogen production from photosynthetic organisms is too low to be biologically viable.

When chlorophyll breaks the water molecule in algae and plants, due to its affinity to light is among 400 and 700 nm, the rest of the luminous energy gets lost, is approximately 80% of energy wasted, unlike melanin in which electromagnetic radiations absorption comprehends the whole visible spectrum y perhaps even more, plus it seems to absorb kinetic energy also. This is quite possible within biology, for it was something to expect that nature could use all kinds of ways to energize all biochemical reactions that give life continuity.

Until now, there hasn't been detected o describe the ability of non plants organisms, both uni and pluricellulars to use electromagnetic spectrum radiant energy to energize one or some of the several reactions that happen inside animal kingdom organisms. These were the reasons why we decided to patent and report any use that comes from the knowledge of melanin as a water electrolyzing element, because due to its affinity with the electromagnetic spectrum – which goes from 200 to 900 nm at least, an maybe more – and due the physiological characteristics of the tissues where this pigment is usually found, where the parameters as at tissue, or arterial, capillary and venous oxygen concentrations are notable. We decided patent and register any use that could come from the fact that, when melanin or melanin are lighted up, we get water molecule's photolysis generating hydrogen and oxygen atoms, and oxygen diatomic mainly and a flow of high energy electrons. Hydrogen molecules move fast, but electrons move to light speed, as well as melanin ability to support and catalyze the opposite reaction, this means join hydrogen and oxygen atoms, forming water molecules as well as electricity. Any use, direct or indirect of this property of melanin, as well as any advance in the study, description, handling and application en several processes in all fields of knowledge (biology, biochemical, physics, physicochemical, etc..) open up a gate of unknown perspectives and, due to the fact that it was unknown that besides plants, and cianobacteria, there were another living being in animal kingdom able to capture sun energy in one way or another, we have patented any use or utility from the knowledge that nature uses melanin to generate energy through water photolysis and uses this energy to the development, impulse or organization of one or multiple biochemical cell processes.

Definitions: the used terminology is according the following definitions:

MELANINS- For the purposes of this work, when we write melanins, we refer to melanins, melanin's precursors, melanin's derivatives, melanins variations and analogues (polyhydroxyindol, eumelanin, pheomelanin, alomelanin, neuromelanin, humic acid, fulerens, graffit, poliindolquinones, acetylene-black, pyrrole-black, indole-black, benzene black, thiophene black, anilina-black, hydrated poliquinone, sepiomelanins, dopa-black, dopamina-black, adrenalina-black, catechol-black, 4 amine catechol-black, (in simple linear chain, aliphatic or aromatic) or its precursors as phenols, aminophenols or diphenols, indolpoliphenols, ciclodopa, DHI and DHICA, quinones, quinones, semiquinones, or hidroquinones, L-tirosine, L-dopamine, morpholine orto benzoquinone, dimorpholine-orto-benzoquinone, morpholine-catechol, ortobenzoquinone, porphirin-black, pterin-black, monochrome-black, nitrogen free precursors, any of the above cited with any particle size, (from 1 angstrom to 3 or 4cm), every compound above, electro-active, suspended, inside gel, inside cytoplasm, inside modified liposome, absorbing ultrasound within a one MHz range, synthetic or natural, whether vegetable, animal or mineral, pure or mixed with organic and inorganic compounds, metallic ions (gadolinium, iron, tin, gold, silver, nickel, copper, erbium, europium, praseodymium, dysprosium, holmium, chromium or manganese, lead selenium, titanium, diverse and special alloys, that is, any metal, known or unknown to this day, etc.) gadolinium is a very effective metal which is integrated inside melanins in ionic shape, or as a particle both in vivo as in vitro, plus drugs, energizing the photoelectrochemical design with electromagnetic radiations, light (synthetic and natural, coherent or not, mono or polichromatic) with a wavelength mainly between 200 and 900 nm, although other wavelengths and another kinds of energy, for instance, kinetic energy, are also efficient if variable, depending on all the conditions (pH, temperature, pressure, combination, doping, etc..)

PHOTOLYSIS: it refers to the water molecule partition into hydrogen and oxygen, a process energized in melanin by visible and invisible spectrum electromagnetic radiation's absorption.

AQUEOUS HUMOR DYNAMICS: it refers to the synthesis, circulating and excreting of aqueous humor inside de eye.

GLAUCOMATOSE OPTIC NEUROPHATY: A mysterious disease that constitutes one of the world's first three causes for blindness, being more frequent wherever the skin is darker, consists in a optic nerve's atrophy which can course or not with a intraocular pressure rising.

ENERGY: defined as the capacity to execute a determined work. Energy can have several shapes: mechanical, thermal, chemical, etc..

MACULAR DEGENERATION ASSOCIATED TO AGE: It's the first cause for blindness in Caucasian countries, consists whether in the slow disappearing of structures in the macular zone, as vessels, melanocytes, photoreceptor cells, or in occasions its awakes by angiogenesis due to prevalent hypoxia, it can be followed by a descend in melanocytes, melanin and blood vessels inside the zone.

DIABETIC RETINOPHATY: first cause for blindness within the 18 to 65 population. Its physiopathogenesis is not completely known, but hyperglycemia seems to be the setting ground for the illness, because it has been proved that, after a week with a rising in the blood's glucose levels, changes in the eye are taking place, for instance, a widening of the vessels, which can be explained by a rising in oxidant phosphorylation substrates, which results in a bigger oxygen demand, this means, a relatively hypoxia gets developed which the organism compensates with vessels' widening.

DETAILED DESCRIPTION: the whole purpose of this book is essentially to describe and thus to make known the important effects (melanin) has upon cellular biology, the fact, unnoticed until now, that at room temperature and using natural or artificial light as its only energy source, melanin, both in vivo as in vitro, have the ability to break the water molecule, obtaining hydrogen and oxygen diatomic, and also can achieve the opposite event, to gather hydrogen and oxygen molecules, obtaining water and as well as high energy electrons and therefore electric current, to achieve the former, nature uses melanin as unsuspected electrolyzing material. These reactions occurring both inside and outside the cell, even in vitro, and can happen with more or less regularity, under a thousand possible stimuli, whether physical, chemical, both external and internal.

In other words, melanin's effects are so important to nature that cells begins to raise its activity to synthesize this molecule when any reaction, either external or internal begins to occur in a more or less repeated way. Inside eukaryotic cell, more accurately in the melanophores, the main material, the indispensable solute for the electrochemical principle works upon are soluble melanin, given its notable ability to capture photons in wavelengths that goes from 200 to 900 nm of the electromagnetic spectrum, but not only from this part of the spectrum for the pigment is sensible to almost all the electromagnetic spectrum, something that is probably made by the peripheral sections of the molecule, both in vivo as in vitro, which is followed by a high energy electrons generation arising from low energy electrons. These high energy electrons go to the compound's free radical centers where they are probably captured by some element, for instance any metal, such as iron, copper or any other, where they are transferred to a primary electron taker, of unknown nature till today, because the union is very complex and it comprehends ionic interactions, depending from the pH. This electrons transference releases energy, which is used to set a proton gradient.

Water and melanin's molecules combination forms what can be called a photo system, which absorbs luminous energy, using it to, at least, two intertwined actions: to remove electrons from water and to generate a protons gradient. Melanin components are in very close contact one with another, which makes a fast energy transference. In 3 picoseconds after being lighted up, melanin's reaction centers responds sending a photo excited electron towards the primary electron taker.

This electrons transference generates a positively charged giver and a negatively charged taker. The formation of two oppositely charged species importance becomes clear when we consider the oxidize-reduction abilities of these two species, since one of them is electron deficient and can take in electrons, which makes it an oxidant agent. In the other side, the other compound has an extra electron that can be easily lost, making it a reducing agent. This event – the forming of an oxidant agent and a reducing agent from light- takes less than a second's billion part and is the first step in water photolysis.

Due to being oppositely charged, this compounds show an obvious attraction to each other. The charge separation is (probably) stabilized by their own movement, at opposite sides of the molecule, being the negative compound the first to give up its electron to a quinone (Q1), and, possible, then this electron is transferred to a second kind of quinone (Q2), which produces a semi reduced shape quinone molecule, that can be strongly tied to melanin's reaction center. With every each single transference, the electron gets nearer to melanin's reaction center. The positively charged part of melanin is reduced, which prepares the reaction center for another photon's absorption. A second photon's absorption sends a second electron alongside this path (negatively charged melanin towards first and second quinone molecule – (Q1 and Q2) -, this second molecule absorbs two electron, and combines with two protons.

Protons used on this reaction could come from melanin molecule itself or from the surrounding water, causing a descending in the photo system's hydrogen ionic and molecular concentration, which helps to the forming of a proton gradient. Theoretically, the reduced quinone molecule dissociates from the reaction center, being replaced by a new quinone molecule. These reactions occurs at room temperature, but could be modified, for instance, by the heat, that can help reactions in one way or another, depending the control we have on the other known and unknown variables, such as pH, magnetic fields, movement, solute and solvent concentrations, intensity of electromagnetic radiations, the latter's nature, melanin's purity, if it's doped or not, gas partial pressures, container's shape, etc.. And the final purpose we want the process to have, according the kind of cell, tissue, or water available. For instance in presence of metal, such as bore, hydrogen works with -1.

Separating the water molecule in hydrogen and oxygen both diatomic is a highly endergonic reaction due to a very stable association between hydrogen and oxygen molecules. The water molecule's separation (in diatomic hydrogen and oxygen) inside the laboratory requires the use of a strong electric current or rise the temperature almost to 2,000 °C. This (water electrifying or photolysis) melanin accomplishes easily both in vivo as in vitro at room temperature, using only energy from visible and invisible light, and not only from a 200 to 900 nm wavelength, whether from a natural or artificial source, coherent or not, focused or disperse, mono or polychromatic.

The estimated redo potential of quinone's oxidized form is approximately +1.1 V, which is strong enough to attract the strongly tied low energy electrons from the water molecule (redo potential +0.82) which separates the molecule in hydrogen and oxygen atom. Water molecule separation by photo pigments is called photolysis. The forming of an oxygen molecule during photolysis, requires, or so is thought, the simultaneous lose of four electrons from two water molecules, according to this reaction:

$$2H_2O \leftrightarrow 2H_2 + O_2 + 4e^-$$

A reaction center can only generate one positive charge, or its oxidant equal, at the time. This problem is hypothetically resolved with the presence of 4 nitrogen atoms inside melanin's reaction center, each one of these transfers only one electron. This nitrogen concentration stores perhaps four positive charges transferring four electrons (one at the time) to the nearest quinone molecules. The electrons transference from the reaction center's nitrogen to the quinone$^+$, is accomplished by the passing through a positively charged quinone leftover. The electron, after is transferred to the quinone $^+$, is able to regenerating quinone in the process, then the pigment is oxidize alone (again into quinone+) after another photon has been absorbed to the photo system. So, the storing of four positive charges (oxidant equivalents) by nitrogen atoms from the reaction center is modified by the successive absorption of four photons by melanin's photo system. Once the four charges have been stored, the oxygen releasing quinone complex can catalyze the removal of 4e$^-$ from 2EbO, forming an O^2 molecule, and regenerating the completely reduced nitrogen store from the reaction center.

Protons produced by photolysis are released where they are integrated into the protons gradient. The photo system must be lighten up several times before there is an O_2 releasing, and hence, of H_2 also, which could be measurable, which indicates the individual photoreaction's effects must be stored before the O_2 gets released.

Quinones are mobile electrons carriers. We must not forget all electrons transfers are exergonic and occurs as electrons get translated to carriers with a growing affinity to electrons (more positive redo potential). Mobile electrons carriers' necessity is clear. Photolysis generated electrons can go to several inorganic takers, which get reduced by it. These electrons paths can lead to (according the used mix's composition) an eventual reduction of the nitrate molecules (NO_3)into ammonia molecules (NH_3), or sulfates into sulfihydriles (SH), both reactions turn inorganic wastes into compounds useful to life. So, sunlight energy can be used not only to reduce carbon most oxidized form (CO_2), but also to reduce the most oxidized forms of nitrogen and sulfur, both in vivo as in vitro, with obvious advantages. The former lines was cited more deeply and focused to the in vitro form in patent solicitude GT/a/2005/0006.

The production of an O_2 molecule requires removing four electrons from two water molecules. Removing these four electrons requires absorbing four photons, one for each electron. On the other hand, there are also evidences of melanin capturing kinetic energy, turning it into useful energy for the cell, perhaps as hydrogen and oxygen, or the contrary, water and electricity. cytosol and cell membrane's composition are important parameters to obtain this particular reaction's outputs, given the fact that electrolytes' presence, both in vivo as in vitro, its nature, using magnetic fields, using kinetic energy besides electromagnetic radiations, the addition of several compounds – doping – (both organic and inorganic, ions, metals, drugs) into the photo system, which initially only consists of water and melanin, plus electrolytes, as well as temperature handling, gases partial pressures' control, the electric current handling, magnetic fields application, pH levels, makes the final destination to recover electrons, protons or oxygen, as well as resulting compounds, according to the medium in which melanin is dissolved composition. So, the heart of an efficient photo electrochemical design is melanin, both in vivo as in vitro. Electron's transference releases energy, which is used to set up a protons gradient.

Proton's movement along the electrons carrying can be compensated by another ions movement, which are present in the intra or extracellular medium. Melanin's electrolyzing properties (among many others) in vivo, can explain the light generated pike, observed in an electro-retinogram, because when we light up melanin, intracellular pH descends so the H_2 and ionic increases, which activates chlorine channels sensible to pH inside the base-lateral cell membrane. Light's pike is a rising in potential, following the fast oscillating through phase FOT, it forms the slower and longer component of direct current electro-retinogram (Kris 1958, Kolder 1959, Kikadawa 1968, Steinberg 1982).It also intervenes in aqueous humor's dynamics, because they can explain congruently why intraocular pressure descends during daytime, despite the 45% rising of aqueous humor during day, both processes during light's presence, at night aqueous humor production also descends 45% and intraocular pressure rises 40%, so we can observer light energizes notably vital processes of aqueous humor dynamics, which can be supported observing that drugs which stimulate melanocytes indirectly (for instance beta blockers) aren't effective by night. Melanin, melanin's precursors, melanin derivatives, melanin variations and analogues, remove electrons from water and generate a protons gradient.

Light's dependent reactions can also supply energy to reduce CO_2 into CH_2O, nitrates to ammonia and sulfates into sulfihydriles, both in vivo as in vitro. Melanin absorbs all spectrums of electromagnetic radiations, even hard and soft ultraviolet, all visible and invisible spectrum and both near and far infrareds (Spicer & Goldberg 1996). It's not too farfetched that it can also absorb energies like kinetic or other wavelengths far away in the electromagnetic spectrum like gravitons. It is very interesting to consider using melanin's photo-electrochemical properties on industrial processes based on biological systems, like generating hydrogen and oxygen or an alternative generation of electric energy.

Chapter 20

Hydrogen

Hydrogen is the lightest element. It is by far the most abundant element in the universe[2] and makes up about 90% of the universe by weight. Hydrogen as water (H_2O) is absolutely essential to life and it is present in all organic compounds. Hydrogen is the lightest gas[3]. Hydrogen has the symbol H, atomic number 1, atomic weight 1.00794, colorless, classified as non-metallic; his name comes from the Greek words "*hydro*" and "*genes*" meaning "water" and "generator".

Uses:

Robert Boyle (1627-1691) English chemist and physicist; published the paper "New experiments touching the relation betwixt flame and air" in 1671 in which he described the reaction between iron filings and dilute acids which results in the evolution of gaseous hydrogen.

However it was only much later that it was recognized as an element by Henry Cavendish (1731- 1810) an English physicist (who also discovered nitrogen) in 1766 when he collected it

---

[2] As also happens in our body.

[3] Note that while normally shown at the top of the Group 1 elements in the periodic table, the term "alkaline metal" refers only to Group 1 elements from lithium onwards.

(Hydrogen) over mercury and described it as "inflammable air from metals". Cavendish described accurately hydrogen´s properties but thought erroneously that the gas originated from the metal rather than from the acid.

In 1839 a British scientist Sir William Robert Grove carried out experiments on electrolysis. He used electricity from the reaction of oxygen with hydrogen. He enclosed platinum strips in separate sealed bottles, ones containing hydrogen and one oxygen. When the containers were immersed in dilute sulphuric acid a current indeed flowed between the two electrodes and water was formed in the gas bottles. He linked several of these devices in series to increase the voltage produced in a gas battery. Later the term fuel cell was used by the chemist Ludwig Mond and Charles Langer.

In the year1932 Dr. Francis Bacon, an engineer at Cambridge University in the UK worked further on designs on Mond and Langer. He replaced the platinum electrodes with the less expensive nickel gauze and substituted the sulphuric acid electrolyte for alkaline potassium hydroxide (less corrosive to the electrodes). This was in essence the first alkaline fuel cell (AFC) and was called the Bacon Cell. It took Bacon another 27 years to demonstrate a machine capable of producing 5 kW of power, enough to power a welding machine. At about the same time, the first fuel cell powered vehicle was demonstrated.

Much later fuel cells were by NASA in the 1960s for the Apollo space missions. Fuel cells have been used for more than 100 missions in NASA spacecraft. Fuel cells are also used in submarines.

Hydrogen: geological information:

H_2 gas is present in the earth´s atmosphere in very small quantities, but is present to a far greater extent chemically bound as water (H_2O) Water is a constituent of many minerals. Hydrogen is the lightest abundant element in the universe, making up about 90 % of the atoms or 75 % of the mass, of the universe. Hydrogen is a major constituent of the sun and most stars. The sun burns by a number of nuclear processes but mainly through the fusion of hydrogen nuclei into helium nuclei. Hydrogen is a major component of the planet Jupiter. In the planet´s interior the pressure is probably so great that solid molecular hydrogen is converted into solid metallic hydrogen.

Biological role of hydrogen:

Abundance of hydrogen in humans[22].

100000000 ppb by weight[4]

620000000 atoms relative to C (1000000) Atoms of the element per billion atoms.

Hydrogen makes up two of the three atoms in water and water is absolutely essential to life. Hydrogen is present in all organic compounds. A form of water in which both hydrogen atoms are replaced by deuterium (^2H, or D) is called heavy water

---

[4] mg per 1000 kg

(D_2O) and is toxic to mammals. Some bacteria are to known to metabolize molecular hydrogen (H_2).

Hydrogen gas is not toxic but is dangerous if mixed with air or oxygen because of the fire and explosion risk. Halides are defined as the combination of the hydrogen with halogens (F, Cl, Br, I and At); with oxygen are known as oxides; compounds where hydrogen is combined with hydrogen are known as hydrides but not necessary behave chemically as hydrides. In compounds of hydrogen (where known), the most common oxidation numbers of hydrogen are: 1, and -1.

Reaction of Hydrogen with air

Hydrogen is a colorless gas, H_2, which is lighter than air. Mixtures of hydrogen gas and air do not react unless ignited with a flame or spark, in which case the result is a fire or explosion with a characteristic reddish flame whose only products are water, H_2O.

$$2H_2 \text{ (g)} + O_2 \text{ (g)} \rightarrow 2H_2O \text{ (l)}$$

Hydrogen´s reaction with water

Hydrogen does not react with water. It does, however, dissolve to the extent of about 0.00160 g kg^{-1} at 20°C (297 K) and 1 atmosphere pressure.

Hydrogen does not react with dilute acids or with dilute bases.

Hydrogen: electro negativities

The most used definition of electro negativity is that element's electro negativity is the power of an atom when in a molecule to attract electron density to it. The electro negativity depends upon a number of factors and in particular as the other atoms in the molecule. The first scale of electro negativity was developed by Linus Pauling and on his scale hydrogen has a value of 2.20 on a scale running from about 0.7 (an estimate for francium) to 2.20 (for hydrogen) to 3.98 (fluorine).

The bond energy in the gaseous diatomic species HH is 435.990 kJ mol^{-1} at 298 K. Generally these data were obtained by spectroscopic or mass spectrometric means. The strongest bond for a homonuclear diatomic species is that of dinitrogen, N_2 (945.33 ± 0.59 kJ mol^{-1})[23].

There are no data about values of lattice energies for any oxides of H.

Hydrogen: physical properties 3

Melting point	14.1 K (-259.14°C) (-434.45 °F)
Boiling point	20.28 K (-252.87 °C) (-400 °F)
Liquid range	6.27 K
Critical temperature	33 K (-240 °C) (-400 °F)
Superconduction temperature	No data K
Thermal conductivity	0.1805 W m^{-1}K^{-1}
Molar volume	11.42 cm^3
Velocity of sound	1270 m s^{-1}
Elastic properties	No data
Hardnesses	No data

Electrical resistivity	No data
Reflectivity	No data
Refractive index	1.000132 (gas; liquid 1.12) (no units)

Hydrogen: crystal structure

Space group: P63/mmc (space group number: 194). Structure: hcp (hexagonal close-packed). Cell parameters: a: 470 pm, b: 470 pm, c: 340 pm, α: 90.00°, β: 90.00°.

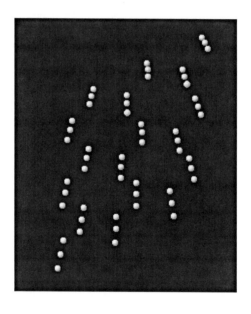

Each sphere in the picture[24] is a molecule of hydrogen (H_2).

Thermochemistry:

http://www.webelements.com/hydrogen/crystal_structure_pdb.html

Diatomic molecules

Diatomic molecules are molecules composed only of two atoms, of either the same or different chemical elements.

Common diatomic molecules are hydrogen (H_2), nitrogen (N_2), oxygen (O_2) and carbon monoxide (CO).

At room temperature seven elements exist as homonuclear diatomic molecules: H_2, N_2, O_2, F_2, Cl_2, Br_2, and I_2. Many elements aside from these form diatomic molecules when evaporated. The noble gases do not form diatomic molecules.

About 99 % of the Earth atmosphere is composed of diatomic molecules. The natural abundance of diatomic hydrogen (H_2) in Earth atmosphere is only on the order of parts per million, but H_2 is the most abundant diatomic molecule in nature. The interstellar medium is dominated by hydrogen atoms.

The bond in a homonuclear diatomic molecule is non-polar. Diatomic molecules cannot have any geometry but linear, as any two points always lie in line. This is the simplest spatial arrangement of atoms after the sphericity of single atoms.

Energy states of real diatomic molecules

For any real molecule, absolute separation of the different motions is seldom encountered since molecules are simultaneously undergoing rotation and vibration. Chemical bonds are neither rigid nor perfect harmonic oscillators, however, and all molecules in a given collection do not possess identical rotational, vibrational, and electronic energies but will be distributed among the available energy states in accordance with the principle known as the Boltzmann distribution.

As molecule undergoes vibrational motion, the bond length will oscillate about an average of internuclear separation. If the oscillation is harmonic, this average value will not change as the vibrational state of the molecule changes; however, for real molecules the oscillation are anharmonic. The potential for the oscillation of a molecule is the electronic energy plotted as a function of internuclear separation. Owing to the fact that this curve is nonparabolic, the oscillations are anharmonic and the energy levels are perturbed.

Since the moment of inertia depends on the internuclear separation, each different vibrational state will possess a different value of $I = ur^2$ and therefore will exhibit a different rotational spectrum. The nonrigidity of the chemical bond in the molecule as it goes to higher rotational states leads to centrifugal distortion; in diatomic molecules this results in the stretching of the bonds, which increases the moment of inertia.

The total of these effects can be expressed in the form of an expanded energy expression for the rotational and vibrational energies of the diatomic molecule. A molecule in a given electronic state will simultaneously possess discrete amounts of rotational and vibrational energies. For a collection of molecules they will be spread out into a large number of rotational and vibrational energy states so any electronic state change (electronic transition) will be accompanied by changes in both rotational and vibrational energies in accordance with the proper selection rules. Thus any observed electronic transition will consist of a large number of closely spaced members owing to the vibrational and rotational energy changes.

Characteristic Temperatures

Characteristic temperature of vibration of diatomic molecules		Characteristic temperature of rotation of diatomic molecules	
Substance	$\theta_{vib}(K)$	Substance	$\theta_{rot}(K)$
H_2	6140	H_2	85.4
O_2	2239	O_2	2.1
N_2	3352	N_2	2.9
HCl	4150	HCl	15.2
CO	3080	CO	2.8
NO	2690	NO	2.4
Cl_2	810	Cl_2	0.36

6140 K = 5867 °C

85.4 K = -187.6 °C

The rotational frequency of H_2 is quite large, only the first few rotational states are accessible to at 300 K.

The electronic energy levels are generally very widely separated in energy compared to the thermal energy kT at room temperature. In each electronic level, there are several vibrational levels and for each vibrational level, there are several rotational states. This is a simplified and useful model to start with. The total energy is a sum of all these energies and is given by:

$$E_{total} = E_{electronic} + E_{vibrational} + E_{rotational} + E_{translational} + E_{others}$$

Usually a very large number of translational states (10^{20}) for volumes of 1 cm^3 for a typical small molecular mass, this means that such a large number of translational states are accessible or available for occupation by the molecules of a gas.

Recall that the rotational frequency of H_2 is quite large, only the first few rotational states are accessible to at 300 K. Also, very few vibrational states are accessible.

Molecule	g	Bond Length (Å)	$\bar{\omega}$, cm^{-1}	Θ_{vib}, K	$\bar{B}$, cm^{-1}	Θ_{rot}, K	Force constant k (dynes /cm)	D_0 (kcal/mol)
H_2	1	0.7414	4400	6332	60.9	87.6	5.749	103.2
D_2	1	0.7415	3118	4487	30.45	43.8	5.77	104.6
N_2	1	1.097	2358	3393	2.001	2.99	22.94	225.1
O_2	3	1.207	1580	2274	1.446	2.08	11.76	118.0
Cl_2	1	1.987	560	805	0.244	0.351	3.2	57.1
CO	1	1.128	2170	3122	1.931	2.78	19.03	255.8
NO	2	1.15	1890	2719	1.695	2.45	15.7	150.0
HCl	1	1.275	2938	4227	10.44	15.02	4.9	102.2
HI	1	1.609	2270	3266	6.46	9.06	3.0	70.5
Na_2	1	3.096	159	229	0.154	0.221	0.17	17.3
K_2	1	3.979	92.3	133	0.0561	0.081	0.10	11.8

Representative molecular data for a few diatomics[25].

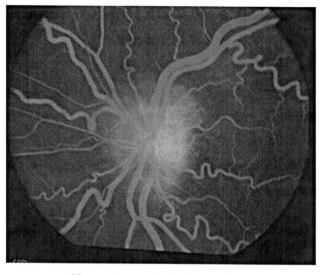

The optic nerve in the live patient.

Chapter 21

Diatomic Oxygen (O_2)

Atomic Number: 8

Mass Number: 15.9994

Electron Configuration: $1s^2 2s^2 2p^4$

Boiling Point: -183°C

Melting Point: -218.4°C

First Ionization Energy: 1314 kJ/mol

Atomic Radius: 74 pm

Ionic Radius: 140 (O^{2-})

Electronegativity: 3.4

As an element, oxygen occurs as a diatomic element in which the oxygen atoms are connected by a double bond.

It has a bond length of 121 pm and a bond energy of 498 kJ.mol^{-1} . This is the form that is used by complex forms of life, such as animals and plants, and is the form that is a major part of the Earth's atmosphere. Dioxygen is the normal is the normal condition of oxygen at room temperature. By comparison, nascent oxygen has only one atom per molecule, and is represented as O or sometimes (O). The parentheses indicate that nascent oxygen does not exist very long under normal conditions. It has a tendency to form Dioxygen.

Oxygen gas is colorless and odorless, but is pale blue in color as a liquid. Oxygen is paramagnetic, that is, they form a magnet in the presence of a magnetic field, a property which can only be explained through molecular orbital theory. Oxygen combines with nearly all other elements on the periodic table. Oxygen will react with most metals to form oxides, with hydrogen to form water, and with nonmetals such as sulfur and fluorine, Oxygen is the second most electronegative element on the periodic table, next to fluorine. In most covalent bonds with oxygen the bonding electrons are shared unequally, resulting in a partial negative charge on the oxygen atom and a partial positive charge on the bonded atom.

When hydrogen and oxygen combine, they give off large amounts of energy. The energy is used to lift the rocket into space.

Oxygen is listed as O_2. The formula of water is H_2O, for every one oxygen atom there is two hydrogen atoms. However, oxygen cannot be listed alone, has to pair up. Therefore a more realistic formula of water should be $2H_2O_2$.

In Molecular Orbital Theory, the bonding between atoms is described as a combination of their atomic orbitals. While the Valence Bond Theory and Lewis Structures sufficiently explain simple models, the Molecular Orbital Theory provides answers to more complex questions. In the Molecular Orbital Theory, the electrons are delocalized (continuously change their positions). Electrons are considered delocalized when they are not assigned to a particular atom or bond (as in the case with Lewis Structures). Instead, the electrons are smeared out across the molecule. The

Molecular Orbital Theory allows one to predict the distribution of electrons in a molecule which in turn can help predict molecular properties such as shape, magnetism, and Bond order.

An oxygen atom is located in the 16th period of the periodic table so its outermost level of electrons (level 2) contains 6 electrons which are two less than eight. So, in order for oxygen to fill its outermost level it needs to gain two electrons, or in this case, form a covalent bond with another oxygen atom this is why an oxygen atom has a subscript of 2.

Because in accordance with the octet rule, the atoms of low atomic number (<20) tend to combine in such a way that they have eight electrons in their valence shells, therefore giving them the same electronic configuration as a noble gas.

In a molecule of oxygen, the two atoms of oxygen are bonded to each other by covalent bonding. The two atoms of oxygen share their two valence electrons and achieve stable electronic configuration.

Almost all molecules encountered in daily life exist in a singlet state, but molecular oxygen is an exception. At room temperature, O_2 exists in a triplet state, which would require the forbidden transition into a single state before a chemical reaction could commence, which makes it kinetically nonreactive despite being thermodynamically a strong oxidant. Photochemical or thermal activation can bring it into singlet state, which is strongly oxidizing also kinetically.

The oxygen content in the body of a living organism is usually highest in the respiratory system, and decreases along any arterial system, peripheral tissues and venous system, respectively. Oxygen content in this sense is often given as the partial pressure, which is the pressure which oxygen would have if it alone occupied the volume. However, distal measures shows that partial pressures of diatomic oxygen, expressed in percent, have no differences in the Sp O_2% between the hand´s fingers and the toe´s fingers.

Principles of Molecular Orbital Theory

1.) Total number of molecular orbits is equal to the total number of atomic orbitals from combining atoms

2.) Bonding molecular orbitals have *less energy* than the constituent atomic orbitals before bonding

3.) Antibonding molecular orbitals have *more energy* than the constituent atomic orbitals before bonding.

4.) Following both the Pauli Exclusion Principle and Hund's rule, electrons fill in orbitals of increasing energy.

5.) Atomic orbitals are best formed when composed of atomic orbitals of like energies.

In molecules, atomic orbitals combine to form molecular orbitals which surround the molecule. Similar to atomic orbitals, molecular orbitals are wave functions giving the probability of finding an electron in certain regions of a molecule. Each molecular orbital can only have 2 electrons, each with an opposite spin. Compared to the original atomic orbitals, a bonding molecular orbital has lower energy and is therefore more stable.

Where the atomic orbitals overlap, there is an increase in electron density and therefore an increase in the intensity of the negative charge. This increase in negative charge causes the nuclei to be drawn closer together. This smaller attraction leads to the higher potential energy. Due to the lower potential energy in molecular bonds than in separate atomic orbitals, it is more energy efficient for the electrons to stay in a molecular bond rather than be pushed back into the 1s orbitals of separate atoms. This is what keeps bonds from breaking apart.

Diatomic oxygen has a Bond Order of 2, and is considered Diradical in accordance with the FMOs. O=O is a stable gas, paramagnetic, Diradical with singlet and triplet states. The Bond Order is calculated as follows:

Bond Order = (electrons in bonding MOs – electrons in Antibonding

$$2$$

- There are two unpaired electrons, so molecule is paramagnetic.

The main differences between the MO diagrams are the relative energies of the orbitals.

CHAPTER 22

Properties of Water

Water is the main solvent in the body. It has several important properties, for instance:

- A very high molar concentration
- A large dielectric constant
- A very small dissociation constant

Its concentration in biological systems is very high: 5.5 Molar at 37°C. This is almost 400 times the concentration of the next most concentrated substance in the body (Na^+) which significance is that water provides an inexhaustible supply of hydrogen ions for the body, however, given the very small dissociation constant this fact cannot be true, besides that water requires energy to act as hydrogen donor because the separation of the hydrogen from oxygen needs unavoidable energy, otherwise will happen randomly which are not adequately in time and form for a biological system.

Calculations of Water Concentration

Molecular weight of H_2O = (1 + 1 + 16) = 18, so one mole is 18 grams.

One ml of liquid H_2O weights about 1 gram, so 1 liter weighs 1,000 grams.

$[H_2O]$ = 1000/18 = 55.5 moles/liter

However the H_2O formula it is not compatible with the most common form of elemental oxygen that is O_2, therefore a more congruous formula is as follows: $2H_2O_2$

Molecular weight of $2H_2O_2 = (1 + 1 + 1 + 1 + 16 + 16) = 36$, so one mole is 36 grams.

One ml of liquid $2H_2O_2$ weighs about 1 gram, so 1 liter weighs 1000 grams.

$[2H_2O_2] = 1000/36 = 27.777777$ moles/liter

The large dielectric constant means that substances whose molecules contain ionic bonds will tend to dissociate in water yielding solutions containing ions. This occurs because water as a solvent opposes the electrostatic attraction between positive and negative ions that would prevent ionic substances from dissolving in accordance with Coulomb Law:

$$F = (k. q_1 . q_2) / D . r^2$$

Where F is the force between the two electric charges q_1 and q_2 at a distance r apart D is the dielectric constant of the solvent.

The large dielectric constant of water means that the force between the ions in a salt is very much reduced permitting the ions to separate. These separated ions become surrounded by the oppositely charge ends of the water dipoles and become hydrated. This ordering tends to be counteracted by the random thermal motions of the molecules (Brownian movement). Water molecules are always associated with each other through as many as four hydrogen bonds and this ordering of the structure of water greatly resists the random thermal motions. Indeed it is this hydrogen bonding which is responsible for its large dielectric constant.

Water itself dissociates into ions but the dissociation constant is very small (Kw = 4.3 x 10 $^{-16}$ mmol/l). The apparent paradox here is that though this is incredibly small, it has an extremely large effect in biological systems, traditionally explained due to the dissociation produces protons (H^+) that are very reactive and have a biologic importance out of proportion. Furthermore the half life of these protons (H+) are in the order of 10 $^{-16}$ of second. Therefore the paradox cannot be solved. However if take in account that the universal photosynthesis system composed by Light/ Melanin/ Water release H_2 or diatomic hydrogen and molecular oxygen (O_2) a highly stable molecule, in a incessantly manner, then the apparent paradox is less complex, because diatomic hydrogen doesn´t combine with water and it is the molecule which the real large effect in biological systems. By other side when both these molecules are recombined (H_2 and O_2) then a 4 high energy electrons are released. The products of the water dissociation and back-bonding (reformation) are described as separated compounds; however these are mental structures only, in reality they cannot exist as isolated entities.

A major characteristic of the saline water habitat is that all organisms exist in a state of osmotic disequilibrium. Energy is therefore required constant and incessantly to maintain the body fluids at the correct osmotic concentrations.

The limited oxygen solubility and its slow diffusion in water make breathing hard work. If we place melanin inside or covered by semi-permeable membranes, immersed in water, which is: melanin will be separated from water by the semi-permeable membrane which only allows passage of water, the levels of O_2 dissolved in water will be significantly elevated in comparison with water alone.

The Unicity of Water

In the very popular book Theology in Science, by the ubiquitous Dr. Brewer, first appeared in 1860 but was still thriving 30 years later. He expounds the large-scale phenomena of evaporation and circulation of water and draws the usual conclusions. Brewer also deals with two laws: the expansion on cooling and the phenomenon that arise from a high specific heat *("water can receive or lose heat without showing it"). Steam is capable of "holding so large a quantity of heat". "When water is heated through a given number of degrees it absorbs twice as much heat as any other substance". Water "is the great cleansing agent of the world".*

By other side, the anomalous density of water just above its freezing temperature (4°C) and its familiar effects on lakes, fishes, and vegetation are well known facts since centuries ago.

James M. Wilson[26], mathematics master at Rugby School. Ordained in 1879, and headmaster of Clifton College (from 1890) Like all his predecessors, he was deeply impressed by the atmospheric and meteorological effects of water circulation. But unlike them, he tried to focus more strongly on the anomalous properties of water, *"in one property of expansion, water is unique, absolutely unique*". This is its expansion on cooling below 4°C. The consequences of ice floating on water for marine life and the structures of rocks and soil are clearly displayed.

Wilson goes on to admire the liquidity of water, noting that its existence in that state depends on many parameters and that *"these elements are adjusted to one another in such a manner as to produce the actual result we see"*. *Water is once more unique. Its specific heat, as this property is called, is far greater than that of all other bodies, solid or liquid.* The very high specific heat is used to account for the large-scale transference of heat in the oceans, as in the Gulf Stream.

The high latent heat of fusion of ice prevents too rapid a thaw with vast flooding in consequence[27]. When it comes to converting liquid water to vapor by the expenditure of heat, no other liquid *"approaches water in the quantity it requires"*.

And this *"extraordinary latent heat of vapour"* means that vast stores of heat are transferable across the globe, to be discharged when most needed. One remarkable property of water is the power of absorption of radiant heat, or its ability to trap heat.

John Tyndall wrote: *Aqueous vapour will not let heat radiated from the earth pass through. It absorbs it; it serves as a covering, as a light but warm blanket to the earth at night . . . this singular and beautiful property of water.* This must be one of the earliest descriptions of what we now call the "greenhouse effect" of water vapor. It is, in fact, at least as significant as that of CO_2.

Wilson's final point of anomaly is the solvent ability of water[28]. *"Its powers of dissolving salts, and of fertilizing the soil, are perhaps the most extraordinary"*

Each property of the qualities of water has been skillfully adjusted to some special end, as it does in the harmonious working of all the separate details, and it is only the limitations of our knowledge and faculties which weaken the impression on our minds[29]. *"The grand result is a harmonious system"*, *"water is the most universal solvent known"*

Lawrence Henderson[30], in his book, *The Fitness of the environment*, wrote: Darwinian fitness is compounded of a mutual relationship between the organism and the environment. Besides, he emphasizes specific heat, showing how it may be derived and enumerating some of the consequences of the unusually high specific heat for water. The effects include a relatively constant temperature in streams, lakes, and oceans, and also a moderation of the summer and winter temperatures on earth.

Marine organisms are able to move about, and animals like man maintain constancy of body temperature without too much difficulty since water is a high proportion of their bodies. Moving on to the high latent heat of water, Lawrence Henderson observes that the large amount of heat required melting ice ensures the temperature of the underlying water is fairly constant.

Therefore is not surprising that the latent heat of evaporation is by far the highest known. Meteorologists are considering the latent heat of evaporation as one of the most important regulatory factors of weather.

First, it operates powerfully to equalize and to moderate the temperature of the earth; secondly, it makes possible very effective regulation of the temperature of the living organism; and thirdly, it favors the meteorological cycle. All of these effects are true maxima, for no other substance can in this respect compare with water.

In referring to the great solvent powers of water we could say that more than 50 individual substances known to be present in urine. By other hand, the great surface tension of water leads to its rise in capillary systems and to complex phenomena of adsorption, important for its passage through soil and the processes of chemical physiology.

In 1934, a book appeared with the title The Great Design. In which E. Armstrong said that water is a material of intense activity, and all vital changes take place when water decomposes into hydrogen and hydroxyl, but at light of the discovery of human photosynthesis we could say: *"when water decomposes into diatomic hydrogen and oxygen."* That is the very beginning of the processes leading to the formation of sugar in and other organic molecules in humans or mammals.

Water is the buffer to receive the shock of the impinging light waves and transmit to us our share of solar energy, but more exactly we could say: Melanin and water are the buffer to receive the shock of the impinging light waves...

In 1938, the medical writer, W. O. Greenwood, in a book called Biology and Christian Belief wrote: *"of all the thousands of possible liquids other than water, there is not one that would have the faintest possibility of supporting life."*

The biochemist George Wald commented in 1958 that *"we now believe that life, being part of the order of nature, must arise inevitably wherever it can, given enough time"*

We now know that without this hydrogen bond[31], water would boil at 200 K rather than 373 K, the DNA helix would not exist, and life as we know it would be totally impossible.

It is of course true that hydrogen bonds are also found elsewhere, as in the ferociously reactive hydrogen fluoride and in liquid ammonia (although that substance is a gas at ordinary temperatures and pressures). Other compounds having an –O-H bond also exhibit hydrogen bonding but to nothing like the extent of water. Nor do they possess most of its other highly unusual properties. (Michael Denton[32]).

Denton stated also that: *"the properties of water in themselves provide as much evidence as physics and cosmology in support of the proposition that the laws of nature are specifically arranged for carbon-based life".*

A further characteristic of water is the exceptionally low viscosity that seems to be at the right level for supporting life processes, not least those depending on diffusion. Another was proton conductance, highly important in photosynthesis and oxidative phosphorylation.

Denton conclusion is awe-inspiring: *"There is indeed no other candidate fluid which is remotely competitive with water as the medium for carbon-based life. If water did not exist, it would have to be invented. Without the long chain of vital coincidences in the physical and chemical properties of water, carbon-based life could not exist in any form remotely comparable with that which exists on earth. And we, as intelligent carbon-based life forms, would almost certainly not be here to wonder at the life-giving properties of this vital fluid."* [33]

Everything in nature was for the best. The environmental crises of pollution, climate change, extinction of species, and rape of the earth must be faced in our own generation.

All this may well be very interesting, but it is not chemistry, nor does it address the question as to why water should be so necessary to the life, and the photo-system Light/ Melanin/ Water gives the answer.

Four billion years of evolution are beyond our understanding but the very beginning of Life is at least at our reach.

Water is an endless repetition of similar parts. Wallace, since 1903, wrote: *"very similar, if not identical, conditions must prevail wherever organic life is or can be developed".*

The expression of Henderson about the relation between life and the environment presents itself as an unexplained phenomenon, it´s now different with the hitherto unsuspected capacity of melanin to split and reform the water molecule since living involves a constant interplay of matter and energy.

Properties of water, such as density, hydrodynamics, viscosity, diffusion, optical, acoustic, thermal and electrical properties, and surface tension, are all utilized by many species to facilitate motion, awareness of, and defense against predators, however are of secondary importance next to the energy production.

Outstanding among mysteries is the management of water movement in plants; however, the unsuspected energy production by lignin, the melanin of plants, will help to resolve the mystery.

Water:

Water is a strange and eccentric liquid, beside to be ubiquitous and commonplace; water is colorless, transparent, and tasteless. The ice floats because of the hydrogen bonding imposing a perfect tetrahedrally coordinated network[34], linking them into six-membered rings with much empty space between the molecules. This is the best known of what are widely seen as a long list of curiosities.

Water is H2O: two hydrogen's, each one attached to a central oxygen; the basic nuclear geometry of the molecule, with the bonded O–H distance of just less than 1 Å and H–O–H angle of about 104.5°. The water molecule is not a static entity; the molecule is in continuous internal motion, with the constituent atoms vibrating against each other.

Interactions between water molecules through the well-known hydrogen bond perturb symmetric and antisymmetric O–H stretch vibrations with respective frequencies of 453 meV and 466 meV, and a bending mode with a frequency of 198 meV from their isolated-molecule values.

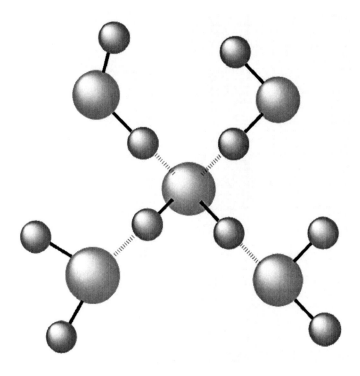

The molecule is made up from one heavy atom (oxygen) and two light atoms (the hydrogens). Therefore there is possibility that quantum effects might be relevant in some processes that involve water. Even at absolute zero, there is still zero-point motion.

The molecular chemistry depends on the electrons. And the water molecule has three vibrating nuclei; therefore we need to consider its electronic structure. Water has a high dielectric constant whose explanation is poorly understood.

A charge distribution is not only to do with the electrostatic attractions between molecules; it is also relevant to molecular repulsion. Water is a small molecule.

There are always some H+ and OH– ions—or rather, their hydrated equivalents such as H3O+ or H5O2+—in liquid water. Not only is molecular diffusion unexpectedly high in liquid water, but so also is the conduction of an excess proton. Water has a relatively thermally rigid and resilient network structure, yet one in which molecular motion is easily possible.

The strength of the hydrogen bond interaction between two water molecules has value of around 20 kJ mol–1 (ca. 5 kcal mol–1) that is about 10 times a typical thermal fluctuation at room temperature ($kT_{ambient}$ ≈ 2.5 kJ mol–1) Being significantly stronger than a typical van der Waals interaction.

Even the simplest molecular reorientation would be expected to require the breaking and reforming of at least two hydrogen bonds.

The maximum density of water is at 4°C and its unusually high thermal capacity is also familiar anomalies. Both the melting and boiling points of water are unexpectedly high when it is placed in the sequence of group VI hydrides.

If water is super-cooled very rapid the water fails to crystallize and becomes a glass[35]. This is very interesting because in its high density form it is the most abundant ice in the universe, where it is found as a frost on interstellar grains.

This is not the only regime in which water becomes amorphous. When cooled, water becomes exceptionally compressible, to extent that, if this property did not exist, the oceans would be about 40 m higher. The diffusivity of water is also anomalous, with the observation that initially as the pressure increases so does diffusivity.

Water is very biophilic, in spite that the intrinsic peculiarities of water are best understood by physicists. Biological functions are extremely sensitive to water properties. Even small doses of heavy water (D_2O) are known in at least some cases to be highly toxic.

The collective properties of water are central to both cellular mechanisms and macroscopic properties that range from its viscosity to planetary habitability. The organisms are largely composed of water under standard pressure and temperature[36]. However, even terrestrial life copes with water across a temperature range from well above boiling to below freezing, and ambient pressure from the rarefied heights near the top of the troposphere to the crushing environment of the oceanic trenches and within the Earth´s crust.

At elevated pressures and temperatures, water in some sense becomes less anomalous and less biophilic. The fine tuning of the specific interactions between water and biochemistry is evident. The interrelationships between water and the various organic substrates are so intimate that water itself must be treated as a biomolecule.

Water is an extraordinary solvent, as well as being far more than a solvent and cannot be considered in isolation. Water is the exact and unique substrate for melanin; otherwise its intrinsic property to split and reform the water molecule cannot be expressed.

Water also serves as a ligand. For instance, in both hemoglobin and cytochrome oxidase, the binding and subsequent release of water molecules is critical to their proper function. In the context of protein function the role of water is by no means completely understood.

The hydrophilic and hydrophobic interactions between water and a protein are crucial in successful folding, it is clear that the process is exceptionally finely balanced between competing demands. By one side proteins require stable folds, robust to environmental fluctuations, but by other hand must be flexible enough for allostery and complex interactions within the proteome while simultaneously exhibiting a rich and variable designability within the evolutionary random walk.

The free energy (ΔG) of the folded protein is remarkably low, being equivalent to two to four hydrogen bonds. The process of folding entails a very subtle range of weak, local interactions between molecules in an aqueous medium.

We still do not have a good understanding of water and its interactions with other molecules. The ability of water to exchange entropy with folding protein is one of its more astonishing properties. In spite to be strongly exothermic and therefore highly deleterious to cell function is, in terms of entropy, efficiently compensated.

The bonding strength between water molecules is enormously high, so explaining why it remains as a liquid at a much higher than expected temperature. This strength confers a sort of rigidity. There is little doubt that this combination of properties is important, perhaps critical, in protein function. Therefore the hydrogen bond in water is of fundamental importance.

Density and surface tension of water are properties that appear as relatively insensitive. Different properties exhibit varying degrees of sensitivity to changes in water geometry and hydrogen bond strength.

The hydrolytic activities of water present a severe barrier to the assembly of many molecules abiotically, not least the nucleotides. Alternative views insist that water is far from ideal, and that other fluids are strong candidates for alien life forms. However, water is the exact substrate for the intrinsic property of melanin to split and reform the water molecule. There is no other alternative therefore life would be literally unimaginable without water.

With the main aim to understand the strange dynamic entity we call life, the unraveling of the photo-system composed by Light / Melanin/ Water, arranged in order of abundance in the universe, allows the emergence of a general theory of organization.

The complexity, stability, robustness and sometimes extreme sensitivity of life are explained by the process that we appointed as human photosynthesis. The general prospect is that life is ubiquitous and is based in physicochemical systems derived all of them from the photo-system Light/ Melanin/ Water.

Beside of its close involvement with life, H_2O, mainly in its liquid state, is "strange and eccentric". The molecular properties of water that are of prime importance are: The sp^3 hybridization of the H2O molecular orbitals, which gives rise to an approximately tetrahedral disposition of four possible hydrogen bonds about each central oxygen atom. Water molecule has two positive and two negative charges, therefore its nature is quadrupolar. The low energies of the O-H-O (hydrogen) bonds, in comparison to covalent bonds or ionic interactions, make these bonds susceptible to minor structural distortions and surprisingly these are the only interactions in which water molecules can participate.

Although the above features are also found in other molecules, their combination in one substance is probably unique. The intimate relationship between water and life´s biochemical processes is demonstrated by substituting the deuteron (^2H) for the proton (^1H), substitution that merely a change in the zero-point energy. Even this minor isotopic substitution is toxic to most life forms. Therefore life processes are so sensitively attuned to the O-^1H-O hydrogen bond energies that the substitution by the heavier deuteron alters the kinetics of biochemical reactions. Only the lowest forms of life, in instance, some protozoa, can tolerate the complete deuteration of their constituent biopolymers, but only if brought about in gradual stages. All higher forms of life exhibit signs of enhanced senescence and eventual death.

Water remained an "element" until 1790, when Lavoisier and Priestly demonstrated that water could be decomposed into air (oxygen) and inflammable air (hydrogen). Much later the question of where water existed in the universe was established that most of it is adsorbed on interstellar dust particles that eventually make up the tails of comets.

The Earth water content, estimated as a total of 1.34×10^9 km^3, of which 97 % make up the oceans. The 99.997 % of the remaining fresh-water is locked in the Antarctic ice cap. The 0.003% represents the fresh water immediately accessible for agriculture, domestic, and industrial use. Of this amount, 75 % is used for irrigation purposes.

The hydrological cycle: transpiration/ evaporation/, followed by precipitation; ensures that our surface water is recycled 37 times per year, constituting a vast water purification system. Unfortunately, 75 % of water precipitation falls into oceans and thus becomes largely useless, at least to the terrestrial animal and plant kingdoms, unless the salt is removed.

Within the overall hydrological cycle of transpiration and precipitation, two sub-cycles are between food producers (plants) and consumers (animals). The nitrogenous and phosphorus cycle are of almost equal importance; they can be easily upset where large quantities of fertilizers and/or detergents find their way into water sources, such as rivers and lakes. The imbalances caused between producers and consumers by excesses of nutrients, such as nitrates and/or phosphates, can have disastrous effects appointed as eutrophicacion (overfeeding) and the Lake Erie during the 1950s is sadly a very good example of this.

The interactions between water and molecules that govern life processes can be studied at several levels of increasing complexity. Over the past decades, much has been written about a so-called water structure. Biological and technological phenomena have been ascribed to this ill-defined water structure and to bound water in strongly or weakly way. Such distinctions run counter to the universally accepted laws of physics, which do not even recognize the existence of molecules at all.

Water can interact directly only by hydrogen bonding, either with ions or with molecules that, like water itself, possess proton donor and/or acceptor sites. Ionic hydration can, to some extent, be treated by the laws of electrostatics. Ionic hydration and hydration by direct hydrogen bonding can be treated by classical physical approaches.

Hydrophobic hydration describes the interaction between water and molecules or ions that "hate" water and are incapable of participating in the formation of hydrogen bonds. Such molecules include the noble gases and hydrocarbons; the simplest example is methanol in which the $-OH$ group favors the interaction with water, but the $-CH_3$ group is hydrophobic and is repelled by water.

The balance between hydrophobia and hydrogen bonding as to which effect will predominate. In the case of methanol, the $-OH$ group wins, making the alcohol completely miscible with water. Molecules forming the cell membranes contain long alkyl chains and only one single polar head group. On balance they are therefore insoluble in water.

Folded, native protein structures are maintained by many stabilizing intrapeptide ionic and hydrogen bond interactions. These are, however, counterbalanced by destabilizing hydrophobic interactions between alkyl residues and water. The net stability margin of a protein in its active state rarely exceeds 50 kJ/mol, an energy equivalent to only three hydrogen bonds in a structure that contains many hydrogen bonds.

For a globular protein molecule to form a biologically active structure, it requires a ca. 50% content of (destabilizing) hydrophobic amino acids. It is thus the fine balance, caused by water-promoted interactions that have given us life's workhorses that are responsible for the majority of biochemical functions. However these concepts will change when beside the properties already known of water the energy released symmetrically in all directions are added to the equation.

We could say so far that biochemistry is the chemistry of water but from now in ahead will be the photochemistry of the Light/ Melanin/ Water, arranged in order of abundance in the universe. The versatility of the water is not enough to gives origin to the life. We must keep in mind that water is just a very important component of the universal photosynthesis system Light/ Melanin/ Water but necessarily the life´s origin requires the complete photo-system; each separate component is unable to gives the very first spark of the life at all.

The H_2O molecule acts as proton-transfer medium in four basic types of biochemical reactions: oxidation, reduction, hydrolysis, and condensation. Melanin oxidizes the water and reduces the oxygen.

Below a threshold value, increasing pressure increases water diffusivity; for a normal liquid, the increased crowding results in the opposite behavior. The ensemble of bio-molecular processes requires an aqueous environment and energy.

The molecular weight (MW) of melanin is estimated in millions of Daltons, the MW of water is 18 Daltons; and the photon is massless. Energy can neither be created nor destroyed, and energy, in all of its forms, has mass. Mass also can neither be created nor destroyed, and in all of its forms, has energy. Whenever energy is added to a system, the system gains mass.

Into the cell everything is fine tuning processes. For instance, the native protein is only marginally stable, in order of 10-20 kcal mol $^{-1}$, which amounts to only two to four of the several hundred hydrogen bonds in a typical native protein solvent-system. Lose a small fraction of these hydrogen bonds and the native structure of the protein falls apart.

Each water molecule is capable of donating two hydrogen bonds through its two protons and accepting up to two through its negatively charged region. There is some evidence that a single accepted hydrogen bond is likely to be stronger than a fully four-coordinated molecule.

It is unlikely that the exposed protein surface would present equal numbers of hydrogen bond donors and acceptors to the surrounding solvent[37]; however one of the important properties of the water molecule is the ability to accommodate variable coordination. Water can accommodate itself to varying external hydrogen bonding requirements without its own structure being significantly affected.

Like liquid water, the protein is also held together in significant measure by hydrogen bonds[38]. By other side is interesting that in spite that the water network is relatively rigid allows a rapid molecular diffusion. The diffusivity should not be significantly compromised in the crowded conditions that are found in the cell.

The already known (and unknown) properties of water undoubtedly fit perfectly so the intrinsic ability of melanin to split and reform the water molecule can take place. The variability of the underlying ideal 2:2 donor/acceptor ratio of the water molecule seems to be crucial for melanin internal process and for life processes.

Many important biological processes including water dissociation and reforming; requires the transport of protons to or from a protein active site. Proton mobility in water is anomalously high[39]. Thus, water seems to be a particularly useful medium for facilitating this part of much biologically important process, photosynthesis included.

Melanin optimizes the water molecule properties but the deciphering of how and when will require a lot of work. The fine tuning inner characteristics of the universal photosynthesis system bring us to think that Light/ Melanin/ Water are the only way to generate life in the whole universe, and not only over the earth.

The interesting properties of water individually are demonstrated by other molecules; but only in water itself all these properties are found[40].

Water is the unique and universal matrix for life, it is a substance that actively engages and interacts with bio-molecules in complex, subtle, and essential ways. Therefore we could think that the active volume of molecules are beyond their formal boundary or the van der Waals surface because the shell of water that surrounds them is activated, that means that respond to the presence of the molecule.

The presence of an intruding solute particle modifies water´s behavior and the nature of this response is far from obvious. Water is an extremely good solvent for ions, in part as a result of water´s high dielectric constant. Water, also; is an efficient solvent for bio-molecular polyelectrolytes such as DNA and proteins. The electronic properties of water are the best fitted for the expression of the intrinsic capacity of melanin to oxidize water and reduces hydrogen.

Water molecules will solvate cations by orienting their oxygen molecules toward the ion, whereas they will adopt the opposite configuration for anions. Hydrophobic solutes in water experience a force that causes them to aggregate. It seems clear that this hydrophobic interaction is in some way responsible for several important biological processes[41].

However this single explanation is not enough to understand the aggregation of amphiphilic lipids into bilayer, the burial of hydrophobic residues in protein folding, and the aggregation of proteins subunits into multisubunit quaternary structures. It is possible that the alternate waves of diatomic hydrogen followed by reformed water accompanied by a relatively ordered flow of electrons that melanin releases in form of growing spheres, symmetrically in all directions; are the complimentary mechanisms of the so-called important biological processes.

I have no doubt that water alone has a very different molecular behavior compared with a system where melanin and water are combined. Even more; the electronic characteristics of both compounds change dramatically. It is ironic that water is one of the smallest molecules and melanin is the largest one.

Chapter 23

Hydrogen bond is central in molecular biology:

My discovery of the human photosynthesis process or the intrinsic property of melanin to split and reform the water molecule and therefore producing energy by means of light transduction; was made based mainly in the clinical study that I named "Spectroscopy of Retina and Choroidal Layer in the living patient" that was utilized along twelve years of continued studies concerning the three main causes of blindness in the world. This observation´s method in living patient allowed us to be aware of the interplay between the melanin that is contents in higher amount in the choroid (from the Greek- grape color-) and the retinal and choroidal blood vessels.

Infer the hitherto unsuspected capacity of melanin to produce energy arising from the study of the hydrogen, oxygen, carbon, nitrogen, water; or any biological molecule (homonuclear diatomic molecules, heteronuclear diatomic molecules, diatomic molecules, polyatomic molecules) is not easy in anyway. The proof is that in spite of the work of several extraordinary researchers in the field in relatively recent times in the area of the valence electrons or chemistry; or in the field of the atomic nucleus that are studied by physics; or in the area of molecular biology or medicine itself; all these great researchers' didn´t see or at least couldn´t discern this awe-inspiring property of melanin.

My intention to present in these book brief reviews of related themes is with the main aim to prove that in these also fascinating areas, the present knowledge is not enough to infer the marvelous ability of human body to makes the equivalent to photosynthesis in plants or in analogy the human photosynthesis.

By other side, neither is possible that based in the actual knowledge of physics, chemistry, biology or medicine; denied the human photosynthesis process. Furthermore, the current concepts in those fields of knowledge do not have argument that can threw it out, nor the study of sole properties of the molecules implied in the human photosynthesis process could eventually allow us to predict the intrinsic property of melanin to split and reform the water molecule. Moreover the study of the components of the human photosynthesis system which have been gradually better understood along the last decades, is a knowledge which allow to infer in more or less degree that the energetic of the human photosynthesis process and founding during my study of the three main causes of blindness is congruous and coherent with the laws of physics and chemistry and therefore feasible.

More important I cannot give in this moment convincing proof of my discovery from the point of view of other sciences besides energy and medicine, for instance, the science of the covalent bonds, a concept that is just a mental structure; because in fact bonds are not a separate structure.

The natural abundance of hydrogen (H_2) in the Earth's atmosphere is only on the order of parts per million, but H_2 is, in fact, the most abundant diatomic molecule in nature. The interstellar medium is, indeed, dominated by hydrogen atoms.

Electrical forces that act on positively charged nuclei of various atoms and negatively charge electronic clouds that extend around these nuclei rule chemistry. The three other fundamental forces in physics, namely strong and weak interactions that act on the protons and neutrons of the nuclei and gravity do not play any role in chemistry. The first two are much stronger than electromagnetic forces and consequently correspond to much larger energy level separations than energies due to electromagnetic separations. It implies that in chemistry all nuclear levels are ground state levels, or nuclei are always in their fundamental state. The third fundamental force, gravity, is orders of magnitude too weak to have any detectable influence on electromagnetic levels.

The elementary constituents in chemistry are therefore atoms, made of positively charged nuclei that are always in their ground nuclear state and surrounded by negatively charged electronic clouds. The precise knowledge of the structures of these electronic clouds is the object of the chemistry. Atoms are the simplest arrangement of all these electrons and nuclei, even they are not the most stables one.

Two H-atoms, the simplest atoms made of single protons surrounded by single electrons, are attracted to each other in such a way that their initially separated electronic clouds mix together so as to form a single cloud occupied by both electrons with different spins, which keep the two protons separated by a well-defined distance.

The H2 molecule is more stable by -4.5 eV than the configuration defined by the two far-away non-interacting H-atoms. This electric rearrangement of charges with an appreciable energy gain or enthalpy gain, enthalpy is defined as a measure of the total energy of a thermodynamic system.

Enthalpies of covalent bonds typically fall in the range of about -5 eV. The covalent interactions are short range actions. In the case of the H_2 molecule, for instance, its energy is of the order of -4.5 eV when the distance between the two protons is in the vicinity of 0.8 Å, but it rapidly approaches zero when this distance increases.

Atoms may also undergo other interactions. Charged atoms that have lost or gained one or more electrons are ruled by ionic interaction that that we may occasionally encounter. The magnitudes of the enthalpies of these ionic interactions are comparable to those of covalent interactions. Contrary to covalent interactions, ionic interactions are long-range interactions. Furthermore, ionic interactions are barely directional, contrary to covalent interactions that are strongly directional.

The energies of covalent bonds are smaller than atomic energies and much smaller than nuclear energies. Ejecting an electron from an outer orbital of an atom thus requires about 10 eV, which corresponds to the energy hv of a near UV photon. The inner electrons require some keV to be ejected from their atom orbitals. It corresponds to hv of an X-ray photon having a wavelength of the order of 1 nm. Chemical interactions, with enthalpies of about -5 eV only imply outer electrons of atoms, the much greater energies of the inner electrons being hardly affected by the chemical state. Nuclear energies are still greater in the order of MeV. Ejecting a neutron from an atomic nucleus requires about 10 MeV. A fission reaction requires about 100 MeV. Energies involved in chemical reactions, some eV, are thus clearly much too small to induce transitions from ground state levels of nucleons towards excited states.

Covalent interactions are the origin of the stability of molecules and govern their structures. Molecules are well-defined entities that appear as stable arrangements of atoms at room temperature. When two identical molecules come in close proximity they suffer residual electrostatic interactions called Van der Waals interactions. These are the origin of the condensations of gases into liquids when temperature decreases, except in liquid water; where the condensation is almost entirely due to the hydrogen bond.

Energies of Van der Waals interactions are typically of the order of about 0.01 eV for small molecules, which at least two orders of magnitude smaller than the energies of covalent bonds. This electric dipole-dipole interaction and rapidly falls off with distance, however, in smaller values, it is strongly repulsive, therefore a sphere-like forming. No sphere of any atom of another molecule, also characterized by its own Van der Waals radius, can penetrate this hard sphere. It thus defines the shortest distance at which atoms of various molecules can aggregate. Beyond this distance the interaction between the molecules is attractive but decays rapidly. The Van der Waals radii of hydrogen atom are R_H= 1.2 Å, R_O = 1.5 Å for O-atom, R_N = 1.55 Å for N-atom and R_C = 1.71 Å for C-atom.

Hydrogen bonds:

Between these two electrical interactions –covalent between atoms and Van der Walls between molecules—exists then an intermediate interaction, called the "hydrogen bond". Sometimes abbreviate H-bond. It occurs between a molecular group, most often O-H or N-H which carries an H-atom and exhibits a marked electric dipole moment and the O- or N- atoms of another molecule.

The establishment of an H-bond does not destroy covalent bonds. Therefore H-bonds are most of the time interactions between two molecules that retain their individualities. H-bonds could be classified as intermolecular interactions, but can be happen also inside single molecules and then are called "intra-molecular" H-bonds and do not destroy the covalent bonds of the molecule they are part of.

Only H-atom and its isotopic variations D (deuterium) or T (tritium) establish such H-bonds. The energetic hierarchy of this chemical bonds corresponds to the hierarchy of chemical bonds corresponds to the hierarchy of primary, secondary and tertiary structures of proteins. In opposition to Van der Waals interactions most H-bonds are directional and are collinear in their equilibrium state.

H-bonds have for long been considered as anecdotic interactions. Their particular importance is that much life rests on them. Werner in 1902 called this interaction *Nebenvalenzbindung*, a nearly covalent bond. It was during 1920 that the H-bond was recognized as responsible for the stranger properties of liquid water. Pauling famous *"Nature of Chemical Bond"* was the book that made H-bonds known to chemists. IR spectroscopy since 1936 has been the most precise and sensitive tool to observe H-bonds: they are directional and, consequently, at the origin of organized molecules structures that are crucial in chemistry and biology. During the 1970s, scientists became aware that the dynamical properties of H-bonds might be even more fundamental.

Until the 1990s the ubiquity of H-bonds in our surroundings was clearly appreciated, in particular with the ubiquity of the H_2O molecule and its fundamental role in bio-reactions at the molecular level. These aqueous media have been for long time considered as casual media devoid of any special property and, consequently, of any interest[42]. However, at light of our discovery of the intrinsic property of melanin to split and reform the water molecule, water are media with subtle properties that are crucial for our knowledge of many processes, particularly life processes, but that we are still far from understanding precisely. The road to the discovery of the human photosynthesis is indeed much longer through the precise knowledge of unique properties of the light, melanin and water; than through the study of the human retina in the living patient with spectrometry of retina and choroid.

The life processes are basically made of assemblies of H_2O molecules that have the unique ability to develop a hyper-dense "H-bond network". The poor knowledge we still have of this H-bond network and its reactivity based on transfers of protons with a half life extremely short and of H-atoms has been qualified as scandal[43]. Water molecules appear familiar but still are poorly known. Still the precise knowledge of the dynamical properties of H-bonds is certainly a necessary achievement; however the discovery of an energetic process that meets the requirements to constitute the very first spark of life makes easier the understanding of how life proceeds at the molecular level.

Briefly, there are two types of H-bond, the intermolecular H-bond that are formed between two independent molecules; representing the large majority of them. The second one is the intra-molecular H-bonds where the molecular groups are both part of the same molecule. An intra-molecular H-bond involves a single molecule, whereas an intermolecular H-bond involves two molecules that become independent upon disruption of the H-bond. Intermolecular H-bonds characteristically establish relatively strong interactions between molecules in liquid; and strongly influence the magnitudes of temperature and heat of evaporation of this liquid, notably in the case of water. The intra-molecular H-bonds do not modify the interactions between molecules which most often remain Van der Waals interactions. Intermolecular H-bonds are the origin of deviations from perfect gas law.

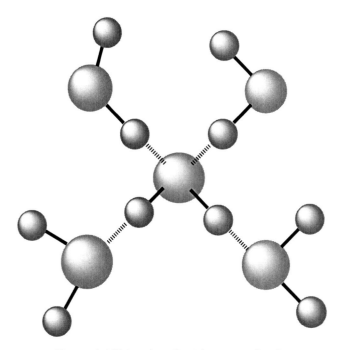

Figure the H-bonds in liquid water molecule.

There exist many intermediates cases, particularly in polymers or macromolecules. They are intra-molecular bonds that suffer only weak constraints from covalent bonds. When two neutral atoms, with their positive nuclei surrounded by spherical electronic clouds approach each other, this results in a strong distortion of the electronic clouds and finally leads to the formation of more stable molecules where the nuclei adopt fixed relative positions and are surrounded by electrons that occupy new orbitals around them. H-bonds cannot be treated in the same way as Van der Waals interactions that only slightly modify the covalent orbitals by mutual polarization and can therefore be handled on the basis of perturbations of orbitals of the non-interacting molecules. H-bonds have stronger effects than covalent bonds, which mean that they require treating the whole H-bonded complex as a super-molecule or molecular complex. It involves a great number of orbitals that rapidly surpasses the possibilities of any computing facility, but if we know where we are going, in this case the understanding of the dissociation and reform of the water molecule, perhaps the necessity of huge computing facility is diminished.

The complexes formed by H-bonds are stable but at the same time they are flexible, evolutive and adaptable, a set of properties that covalent bonds are unable to provide. It is not by chance that water is the perfect substrate for melanin. The answer of the question in regards the behavior of the H-bonds inside the complex Light/Melanin/Water constitutes a formidable task.

The nucleus of the H-atom is composed of a single proton, which of the D-atom (deuterium) has a neutron in addition to the proton. The two nuclei have accordingly a different mass: that of D is twice that of H; however both have the same charge, in consequence the electronic structures of both H- and D-atoms are exactly the same. The origin of H-bonds has a pure electric character, ruled by Coulomb interactions between various charges. When a D-atom replaces the H-atom of an H-bond in physics and chemistry has limited effects, this is not at all so in biology, where such a substitution has nearly always lethal consequences.

The joules (J) used to express energies in the international unit systems are not practical units for molecular systems, in which the electron-volt (eV) unit is used, by mean of equation as following:

$$E = 1eV = 1.60 \ x \ 10^{-19} J$$

The eV is reasonably well adapted to electronic transitions between outer orbitals of atoms or molecules that correspond to energies of the order of a few eV. For H-bonds that are weaker bonds that than covalent bond, other units, such the kilocalorie per mole or kilojoule per mole, are often encountered. The equation is as follows:

$$E = 1 \ eV = 23.04 \ kcal \ mol^{-1} = 96.3 \ kJ mol^{-1}$$

Electromagnetic waves can be characterized by their frequencies, expressed in Hz, or by their wavenumbers $\tilde{v}$ expressed in cm^{-1}. One wave-number is the number of wavelengths in the unit distance $l_0 = 1cm = 10^{-2}$ m and can be defined as follows:

$$\tilde{v} = 1eV = kT_{eV} = h \text{☐}_{eV} = 10^2 hc\tilde{v}_{eV}$$

With T_{eV} = 11.600 K, $\tilde{v}_{eV}$ =8054 cm^{-1} and v_{eV} =241 THz (1 THz = 10^{12} Hz). It defines the correspondence but not equality between eV, K and cm^{-1}. Wavenumbers are equal to frequencies divided by c, the velocity of light.

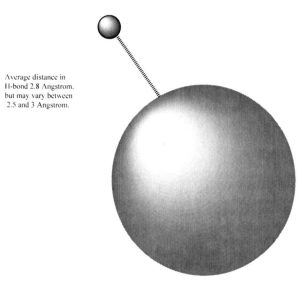

Average distance in H-bond 2.8 Angstrom. but may vary between 2.5 and 3 Angstrom.

Figure: H-bond, black H-atom, red O-atom.

The histograms of a series of several hundreds of such H-bonds around 30 year ago show that the distribution of equilibrium solid angles is a Gaussian distribution centered on 0 and having a width of about 15°. Therefore the H-bond examine is directional and linear within 15°. This directionality is fundamental property of H-bonds. It differentiates H-bonds from Van der Waals interactions that do not exhibit this property, at least at room temperature. It allows for building well-defined supramolecular structures. These supramolecular structures are consequently stable at room temperature but exhibit flexibility, capacities of evolution and adaptability because H-bonds, their cement, can be easily disrupted at the cost of energy. Again, energy, not only form and function needs energy, even the evolutive processes too.

With their directionality ad their enthalpies of about 5-10 kT at room temperatures, H-bonds are the only type of chemical interaction that can offer both stability and flexibility, two properties that are necessary in molecular biology.

kT is the product of the Boltzmann constant, k, and temperature, T. This product is used in physics as a scaling factor for energy values in molecular-scale systems, as the rates and frequencies of many processes and phenomena depend not on their energy alone, instead on the ratio of that energy and kT.

Ordinary H-bonds, and bifurcated H-bonds or even three centre H-bonds, apparently observed by means of crystallography but not confirmed by IR spectroscopy, the most sensitive method to observe H-bonds, had been occasionally used as an explanation of some otherwise poorly understood results, especially in the case of the exceptional physical properties of liquid water.

In cellulose, the H-bond network has such an extension that cellulose is one of the most stable and resistant macromolecule encountered in biology. This fact account to my finding that lignin is the plant´s melanin and is found intertwined with cellulose in the trunk. These kinds of H-bond network require energy to be created, molded, and maintained; and to make their inherent functions as the dynamic of the sap´s plants, and unexplained phenomena so far. My point of view is that lignin, the plant´s melanin; is the source of energy of the plants, as melanin does to human body, and by other side the real human chlorophyll is the hemoglobin, both compounds can dissociate irreversible the water molecule.

Chlorophyll molecule

chlorophyll a R$_1$ = CH$_3$
chlorophyll b R$_1$ = CHO

Chemical Formula: C$_{55}$H$_{71}$MgN$_4$O$_5{}^{2+}$
Exact Mass: 891.53
Molecular Weight: 892.48
m/z: 445.76 (100.0%), 446.26 (73.2%), 446.76 (41.1%), 447.26 (11.2%), 447.27 (4.2%), 447.76 (2.9%), 446.26 (1.5%), 447.77 (1.2%)
Elemental Analysis: C, 74.02; H, 8.02; Mg, 2.72; N, 6.28; O, 8.96

Heme group
(Hemoglobin)

Chemical Formula: C$_{32}$H$_{28}$FeN$_4$O$_2$
Exact Mass: 556.16
Molecular Weight: 556.44
m/z: 556.16 (100.0%), 557.16 (37.3%), 558.16 (7.6%), 554.16 (6.4%), 555.16 (2.3%), 557.15 (1.5%)
Elemental Analysis: C, 69.07; H, 5.07; Fe, 10.04; N, 10.07; O, 5.75

310

H-bond network in cellulose is indeed a crystal, but a crystal that exhibits much more flexibility than ordinary mineral crystals. Protein and DNA are two typical types of stable biomacromolecules where flexibility and the possibilities of evolution are given by H-bonds are of fundamental importance. But remember that the evolution itself is a set of biochemical processes that all of them need a source of reliable and constant energy, as the unique output of the photo system composed by Light/ Melanin/ Water are able to give.

The appearance of several types of secondary structures of proteins that can be found throughout their very great varieties illustrates the subtle properties that H-bonds are able to convey. DNA is a structure of elementary components, the nucleotides, that is sufficiently stable but at the same time sufficiently evolutive that it has been able to keep and develop the memory of life for at least 3.6 billion years of evolution. We could say that DNA, the nucleotides, and thereafter their creation, structure and function, beside the evolutive process needs and indeed needed a lot of energy that undoubtedly the intrinsic property of melanin to incessantly split and reform the water molecule could provide so in ancient times as at present.

Cellulose

It is a particular carbohydrate. Carbohydrates form a simple class of biomacromolecules that contain only C-, O- and H-atoms, with the exclusion of N-atoms, and are directly obtained from atmospheric CO_2 and H_2O molecules during photosynthesis performed by plants. Cellulose is made of elementary repeating bricks that originate from a particular form of glucose, the β-D—glucopyranose linked in 1↔4.

Figure: The β-D—glucopyranose molecule.

β-D—glucopyranose, is a saccharide that has the same general properties of them for which molecules with the same atomic constituents adopting same arrangements but with different stereoconformations have completely different chemical properties. Small differences in these stereoconformations may induce great differences in the possibilities of establishing H-bonds. β-D—glucopyranose, like practically all hexoses found in biology, they adopt the stereostable "chair form".

The energy of light captured during photosynthesis can thus be stored in the form of such polysaccharides that can be later digested, for instance amylose, one of the two main components of starch. This is the way vegetal seeds store reserve of food, and not energy. The melanin content of the seeds is the explanation of long time mystery of their eclosion. When there is enough amount of liquid water, by rain or irrigation, the water content of the seed eventually will be sufficient in a way that the universal photosynthesis system can reach the adequate proportions in order to produce energy in a peak conditions. With the available energy the biological machinery starts to work and the seed thrives.

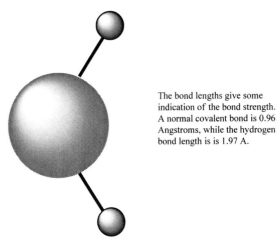

The bond lengths give some indication of the bond strength. A normal covalent bond is 0.96 Angstroms, while the hydrogen bond length is is 1.97 A.

Figure of the water molecule in liquid state.

A free O-H group that does not establish an H-bond has a distance of 0.95 Å, this distance is slightly elongated by the establishment of an H-bond, thus reaching a value of 1 Å for a medium-strength −O-H···O- bond with corresponding O•••O distance of 2.6 Å and reaching a value of 1.1 Å for a strong −O-H...O bond such as that encountered in acid salts.

For biomacromolecules the appearance or disappearance of the supramolecular organization may completely change their properties, making for instance a living protein irreversible becomes a bio-inert polypeptide.

The primary interaction that defines the structure of a molecule or a macromolecule is covalence, the strongest interaction in molecules. H-bonds intervene in their structures only inside the few degrees of freedom that may be left after covalent interactions have been established. In molecules, most of these degrees of freedom left by covalent bonds come from free rotations around bonds such as C-C, or other σ-bonds that are not part of a cyclic structure. They make these macromolecules free to move inside a limited range of various conformations with nearly the same electronic energies. Within this space the possibility of establishing H-bonds strongly favours some conformations that then become more stable than other conformations with no H-bonds. These special H-bonded conformations define the "secondary structure" of these macromolecules. This is the case of cellulose or proteins.

CHAPTER 24

Erythrocytes and Human Photosynthesis

Already 2.4 million of new erythrocytes are produced per second[44]. The cells develop in the bone marrow and circulate for about 100–120 days in the body before their components are recycled by macrophages. Each erythrocyte circulation takes about 20 seconds[45]. Approximately a quarter of the cells in the human body are red blood cells[46]. Channichthyidae are the only known vertebrates without hemoglobin, an oxygen transport protein in the blood. Oxygen is absorbed directly through their scale less skin from the water. It is then dissolved in the plasma and transported throughout the body without the hemoglobin protein. The loss of hemoglobin is not fatal because of the cold environment in which Channichthyidae live. Cold water has much higher dissolved oxygen content than warmer water.

Erythrocytes in mammals are anucleate when mature, meaning that they lack a cell nucleus. When erythrocytes undergo shear stress in constricted vessels, they release ATP which causes the vessel walls to relax and dilate so as to makes easier the blood flow; however vasodilatation requires lessening available energy than the tonic state of vessels, therefore the releases of ATP by erythrocyte is not as a source of energy instead is a way to reduces the energy levels of surrounding tissues, because the breakdown of ATP **absorbs** energy. In general terms it is easier to reduce the energetic level of any chemical or biochemical reaction or set of them than increase it, and the interaction between erythrocyte and surrounding tissues is not an exception.

It has been recently demonstrated that erythrocytes can also synthesize nitric oxide enzymatically, using L-arginine as substrate, just like endothelial cells[47] so erythrocyte requires energy indeed and glucose molecule, by other side, it's not a source of energy instead is a primordial metabolic intermediate, therefore this required energy must come from other source and hemoglobin astonishingly is the explanation (See below).

Red cells have nuclei during early phases of erythropoiesis, presently explanation is that erythrocyte extrudes them during development as they mature in order to provide more space for hemoglobin but could be due to the erythrocyte physiology didn´t support or didn´t requires nucleus or both. In mammals, erythrocytes also lose all other cellular organelles such as their mitochondria, Golgi apparatus and endoplasmic reticulum. As a result of not containing mitochondria, these cells use none of the oxygen they transport; instead they produce the energy carrier ATP by the glycolysis of glucose and lactic acid fermentation on the resulting pyruvate. But this kind of ATP production has yields specially lows so we have in this process two mistakes, 1. ATP is not an energy carrier, only is a metabolic intermediate which absorbs energy when is broken down therefore 2. ATP production in erythrocyte must be considered mainly as an indicator of metabolic activity which by one side reflects synthesis of compounds relative to biomass maintenance and by other side has relationship with the availability of molecules that eventually erythrocyte could need in order to makes and adequate physiologic function in several aspects, and not only in oxygen and CO_2 transportation.

Because of the lack of nuclei and organelles, mature red blood cells do not contain DNA and cannot synthesize any RNA, and consequently cannot divide and have limited repair capabilities[48]. This also entails that no virus can evolve to target mammalian red cells.

As red blood cells contain no nucleus, complex protein biosynthesis is currently assumed to be absent in these cells, although a recent study indicates the presence of all the necessary bio-machinery in the cells to do so (Kabanova 2009).

The blood's red color is due to the spectral properties of the heme iron ions in hemoglobin. Each human red blood cell contains approximately 270 million of these hemoglobin bio-molecules, each carrying four heme groups; hemoglobin comprises about a third of the total cell volume. This protein is responsible for the transport of more than 98% of the oxygen (the remaining oxygen is carried dissolved in the blood plasma). The red blood cells of an average adult human male store collectively about 2.5 grams of iron, representing about 65% of the total iron contained in the body[49].

Hemoglobin is the human chlorophyll in the sense that both molecules have the intrinsic property to absorb determined wavelengths in order to harvest photonic energy and with that energy then the water molecule is dissociated being the main product hydrogen which is the real energy carrier, by other side; diatomic oxygen is a very stable molecule.

Chlorophyll absorbs light blue and red, these are the extremes of visible spectrum and hemoglobin absorbs electromagnetic radiations with a wavelength of less 600 nanometers (yellow, green blue) but reflects totally wavelengths around 650 and 750 nanometers, therefore the color perception of erythrocytes is red. Neither Chlorophyll has a green color nor is hemoglobin red. Both chlorophyll and hemoglobin has the intrinsic property to split and only split the water molecule, the product of real value of both molecules in regards to the water dissociation is diatomic hydrogen, the energy carrier by excellence in the Universe and chlorophyll and hemoglobin expulse oxygen, chlorophyll to the atmosphere and also hemoglobin, even in the first step of this process diatomic oxygen can be dissolved in plasma or captioned by the iron of the hemoglobin but eventually these diatomic oxygen is expelled to the atmosphere through the lungs.

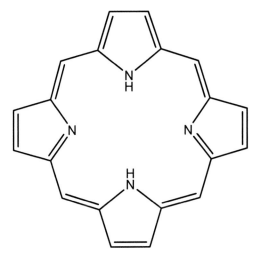

Chemical Formula: $C_{20}H_{14}N_4$
Exact Mass: 310.12
Molecular Weight: 310.35
m/z: 310.12 (100.0%), 311.13 (21.8%), 312.13
(2.3%), 311.12 (1.5%)
Elemental Analysis: C, 77.40; H, 4.55; N, 18.05

Porphyrin a type of pigment found in living things, such as chlorophyll, that makes plants green and hemoglobin which makes blood red.

Erythrocyte has not nucleus or mitochondria, therefore the energy must come from the water dissociation because glucose is just a source of biomass and not from energy.

Chemical Formula: $C_{32}H_{28}FeN_4O_2$
Exact Mass: 556.16
Molecular Weight: 556.44
m/z: 556.16 (100.0%), 557.16 (37.3%), 558.16 (7.6%), 554.16
(6.4%), 555.16 (2.3%), 557.15 (1.5%)
Elemental Analysis: C, 69.07; H, 5.07; Fe, 10.04; N, 10.07; O,
5.75

The heme group: Even though carbon dioxide is carried by hemoglobin, it does not compete with oxygen for the iron-binding positions, but is actually bound to the protein chains of the structure.

Hemoglobin is the iron-containing oxygen-transport metalloprotein in the red blood cells of all vertebrates. Traditionally, the function of hemoglobin in the blood is as oxygen-carrier from the lungs or gills to the tissues of all the body where it releases the oxygen to burn nutrients to provide energy to power functions of the cell, and collects the resultant carbon dioxide to bring it back to the respiratory organs to be dispensed from the organism.

However, the sentence: "oxygen to burn nutrients to provide energy to power functions" should be changed to: oxygen to modify nutrients in order to integrate them to biomass.

In mammals, the protein makes up about 97 % of the red blood cells´ dry content. Hemoglobin has an oxygen binding capacity of 1.34 ml O_2 per gram of hemoglobin, which increases the total blood oxygen capacity seventy-fold compared to dissolved oxygen in blood. The hemoglobin molecule can bind up to four oxygen molecules. The average oxygen consumption in the male is 3.5 liters/minute and 45 ml/kg/min and is properly defined by the Fick equation as follows:

$$VO_2 \text{ max} = Q \, (CaO_2 - CvO_2)$$

Where Q is the cardiac output of the heart, CaO_2 is the arterial oxygen content, and CvO_2 is the venous oxygen content.

This oxygen consumption is relatively small compared with the amount of CO_2 expelled: Twelve breaths per minute and a tidal volume of 500 ml, 8640 liters of air are taken into the lungs every day at rest.

Therefore, to reach a SpO2 of 97 % in blood, our body requires 3.5 liters/minute, but oppositely, to reach a CO_2 blood level of 20 % our body needs 20 liters/minute or more. It has no sense in anyway. Therefore, if we take in account that the oxygen is produced in situ by every single cell in the body by means of the human photosynthesis, that are 100 trillion in average; then the equation becomes coherent. Apparently we need atmospheric oxygen be less picking up and transported to the inner side of body.

In other words: The air that is inhaled is about 21-percent oxygen, and the air that is exhaled is about 15-percent oxygen, so about 5-percent of the volume of air is consumed in each breath and converted to carbon dioxide. The breath rate is 12 to 25 per minute. Size of breath is 500 ml. Percent CO_2 exhaled is 4% so CO_2 per breath is approx 0.04g (2g/L x .04 x .5l).

It is congruous with the existence of human photosynthesis that breathing 100 % oxygen can be harmful, instead to be good for us. The explanation is that the equation is imbalanced as follows:

$$2H_2O \leftrightarrow 2H_2 + O_2 + 4e^-$$

When the oxygen level is increased, several alterations occurs, the fine tuning of the of gas blood transportation for instance is impaired; and more important the turnover rate of cycle is skewed because in presence of high levels of oxygen the equation has a trend to the left and thus the reforming process of water molecule is accelerated, then diatomic hydrogen is consumed and the energetic levels of the cells are turned down with the consequent impairment of cellular functions in an unpredictable manner.

Hemoglobin can be saturated with oxygen molecules (oxyhemoglobin), or desaturated with oxygen molecules (deoxyhemoglobin). The oxyhemoglobin has significantly lower absorption at the 660 nm wavelength than deoxyhemoglobin, while at 940 nm its absorption is slightly higher. The absorption spectrum of hemoglobin shows that the reflection of the molecule appear as designed to diminish at most the presence of red light reflecting it and therefore the absorption of the diatomic oxygen by hemoglobin. At light of our discovery, it is not important the oxygen transportation from the environment to the inner side of body, because diatomic oxygen is produced in situ by every single cell of the organism, in contrast, the role of the oxygen is makes easier the CO_2 transportation, a highly toxic compound. That is the reason of the existence of the carbonic anhydrase, which enhances 10 000 times the dissolution of the CO_2 in the blood. If oxygen were most important, perhaps must be an enzyme to facilitate the oxygen captioning and transportation.

Now we have at least two previously unknown sites of oxygen production: every single cell of the human organism in which interior melanin is synthesized and by other side hemoglobin containing cells: erythrocytes. This unsuspected production of diatomic oxygen allows the perception that gas interchange in the blood stream is more complex that seems until today. The challenge of the gas transportation in regards to oxygen is huge, because diatomic oxygen is produced inside the body and is not only transported from atmosphere to the blood stream. Taking in account that gas exchange is by passive diffusion, this means that there is not energy uses; and then the blood stream receive diatomic oxygen from at least three sources: the cell itself, 100 trillion in average human body; the second source is the hemoglobin contained in each erythrocyte; and finally from the lungs.

Therefore the SpO_2% must be rethinking, but these three unexpected sources of diatomic oxygen explain very well why the SpO_2% is the same in spite the distance from the heart and lungs. The measure of SpO_2% shows similar ciphers in the four extremities, around 98 % in normal patients; which is incongruously with the mechanism of simple diffusion, because as far the lungs are, SpO_2% number should be diminished, but the oppose is thru. SpO_2% are amazingly constant when theoretically must have a similar behavior to $SpCO_2$%, in other words the $SpCO_2$% increase as the lungs are far away, and the SpO_2% must shows an decrease as the distance between the lungs and the measured tissue is increased that is proportionally inverted.

Remember that diatomic oxygen and CO_2 are moving through the organism mainly by simple diffusion (no energy has expensed).

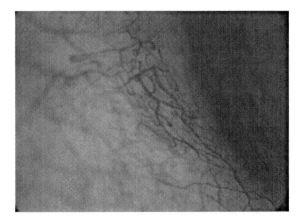

The blood vessels in the conjunctiva have red color when is illuminated with poly-chromatic (white) light, therefore the only wavelength that is reflected is around red light (650 nanometers)

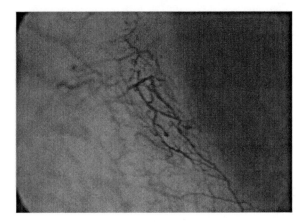

The same tissue when is illuminated with monochromatic light of around 450 nm (green color), then is completely absorbed by hemoglobin so the image is dark, because there isn´t significant light reflection by the erythrocytes' hemoglobin. The wall of the blood vessels are very thin, therefore its effect in light reflection is negligible.

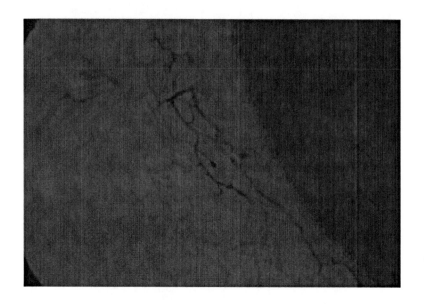

The same conjunctiva tissue but now illuminated with a wavelength closest to 300 nanometers (blue light). The blood vessels image is dark, therefore hemoglobin of the erythrocyte is strongly absorbing this wavelength and there isn´t light reflection.

Are there photo reduction effects of blue light in erythrocyte?[50],[51]

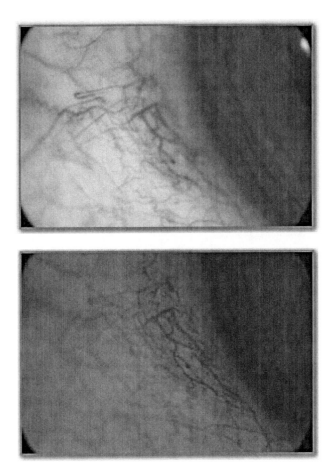

The red color of hemoglobin observed under White light illumination.

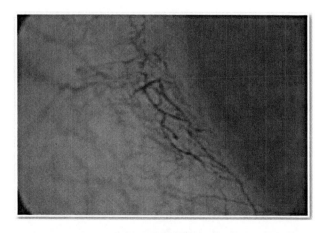

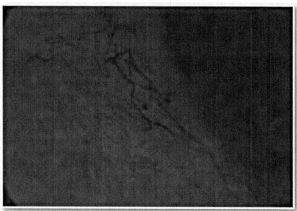

With others wavelengths hemoglobin appears dark.

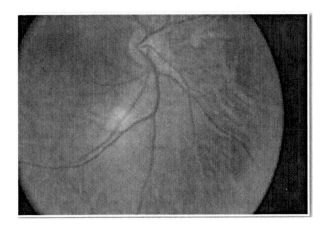

The Optic disc, macular and choroid layer of nasal side,
illuminated with White light.

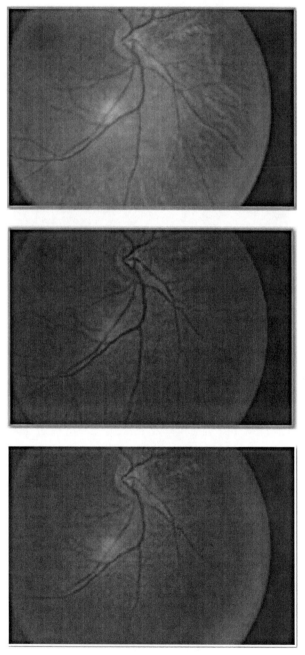

The same area under different monochromatic lights.

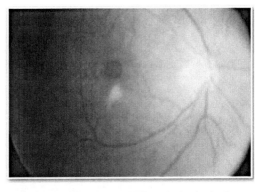

Macular hole, observed with White light (polychromatic).

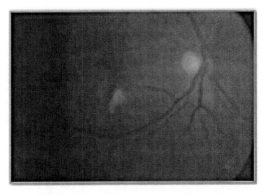

The same case observed with Green monochromatic light.

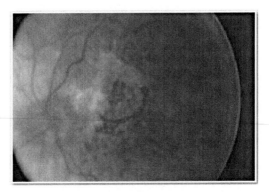

Age-related macular disease photographed with
polychromatic (White) light.

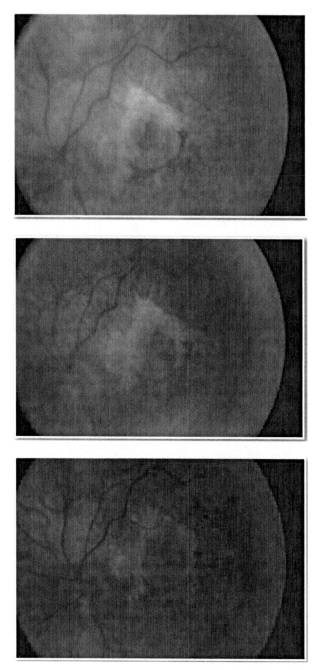

Age-related macular disease, observed with different wavelengths.

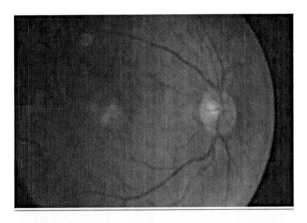

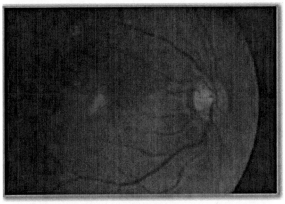

The excavation of the Optic Nerve can be evaluated through comparison of the image taken with different wavelengths.

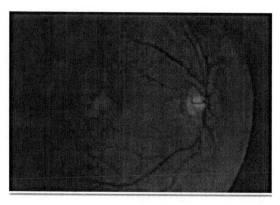

The pathology of the optic nerve modifies the absorption of the different wavelengths and therefore the images are different.

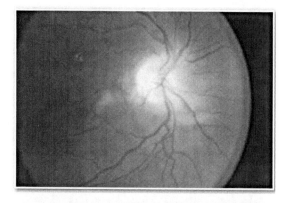

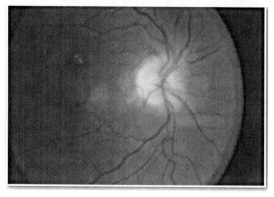

Age-related macular disease studied with different wavelengths.

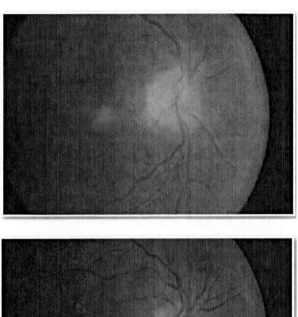

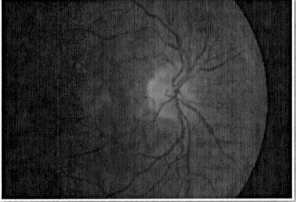

Age-related macular disease studied with different wavelengths.

Chapter 25

African Americans have lower serum 25-hydroxyvitamin D concentrations and a lower risk of fragility fractures than do other populations[52]. Researchers are actively trying to explain it several decades ago. The protective factors against fracture already described in the literature are higher peak bone mass, increased obesity rates (?); greater muscle mass; lower bone turnover rates; and advantageous femur geometry. Bone histomorphometry in young adults shows longer periods of bone formation.

Interesting African Americans fall as frequently as do whites. African American girls accrue more calcium than do whites during adolescence as a result of increased calcium absorption and superior renal calcium conservation. In adulthood, higher parathyroid hormone concentrations do not result in increased bone loss in African American because an apparent skeletal resistance to parathyroid hormone; and their superior renal conservation of calcium persists.

These until now unexplained advantages diminish in the elderly, in who further increases in parathyroid hormone results in increased bone turnover and bone loss. At all ages, African Americans have lower serum 25(OH)D concentrations than do white. Therefore a serum low level of 25(OH)D is undoubtedly a biomarker of good levels of human photosynthesis.

Classically, the explanation is that increased melanin pigmentation interferes with the absorption of ultraviolet B light and therefore with the formation of Vitamin D in the skin.

CO_2, Life and Human Photosynthesis.

In the course of a day a man of average size produces, as a result of his active metabolism, nearly two pounds of carbon dioxide. All this must be rapidly removed from the body. It is not simply the solubility of the CO_2 in water that matters, but also the interaction with the water to form an acidic ion H^+ and a bicarbonate base, HCO^-.

The ability of oceans to dissolve carbon dioxide and to deposit it in the form of limestone is a great boon. If the oceans had not absorbed the gas, it would have remained in the atmosphere. Not only is the solubility of carbon dioxide critical for human life, but so is its easy release back into gaseous form.

In our bodies, carbon dioxide is a principal waste product from the metabolism of carbohydrates, the building blocks source that keeps us functioning and maintains our shape. To eliminate the waste the blood cells carry it to the lungs, where a pressure difference allows the dissolved CO_2 to be released.

The low solubility of carbon dioxide gas in water may help recycle the material; a gas can easily move from place to place. Carbon dioxide is a gas with poor solubility in water; this permits it to be distributed easily on a planetary scale. But water and carbon dioxide prove to be a problematic pair.

The carbon of carbon dioxide is a good electrophilic center. But carbon dioxide itself is poorly soluble in water (0.88 v/v at 293 K and 1 atm). Therefore, at pH 7, carbon dioxide is present primarily in the form of the bicarbonate anion. Bicarbonate, however, has its electrophilic center shielding by the anionic carboxylate group. This means that bicarbonate is intrinsically unreactive as an electrophilic. Thus, the metabolism of carbon dioxide is caught in a conundrum. The reactive form is insoluble; the soluble form is unreactive.

The problematic reactivity of carbon dioxide competes with the problematic reactivity of dioxygen. Even in highly advanced plants, a sizable fraction of the substrate intended to capture carbon dioxide is destroyed through reaction with dioxygen (Ogren and Bowes, 1972).

If we aren´t aware of the intrinsic property of melanin to split the water molecule and therefore of the existence of the human photosynthesis, one comes to realize that the reason why we think that water is optimally suited for life is because it is the solvent that supports the life familiar to us, and for no other reason.

Chemical reactions can take place in the gas and solid phases, but each of these has disadvantages relative to the liquid phase. For example, a hypothetical life form might reside in solids living in deeply frozen water, obtaining energy occasionally from the trail of free radicals left behind by ionizing radiation, and carrying out only a few metabolic transformations per millennium.

Liquid ammonia [NH_3] is a possible solvent for life, and is analogous to water in many of its properties. Ammonia, like water, dissolves many organic compounds, including many poly-electrolytes. Preparative organic reactions are done in ammonia in the laboratory. Ammonia, like water, is liquid over a wide range of temperatures (195–240 K at 1 atm). The liquid range is even broader at higher pressure. For example, when the pressure is 60 times that of the terrain atmosphere at sea level, ammonia is liquid from 196 to 371 K. Further, liquid ammonia may be abundant in the solar system. A large amount of the inventory of liquid ammonia in the solar system exists, for example, in clouds in the Jovian atmosphere. As in water, hydrophobic phase separation is possible in ammonia, although at lower temperatures. For example, Brunner reported that liquid ammonia and hydrocarbons form two phases, where the hydrocarbon chain contains from 1 to 36 CH2 units (Brunner, 1988). Different hydrocarbons become miscible with ammonia at different temperatures and pressures. Thus, phase separation useful for isolation would be conceivable in liquid ammonia at temperatures well below its boiling point at standard pressures. Nevertheless, the increased ability of ammonia to dissolve hydrophobic organic molecules (again compared with water) suggests an increased difficulty in using the hydrophobic effect to generate compartmentalization in ammonia, relative to water. The greater basicity of liquid ammonia must also be considered. The species that serve as acid and base in pure water are H3O+ and HO-. In ammonia, NH4+ and NH2- are the acid and base, respectively. H_3O^+, with a pKa of –1.7, is ca. 11 orders of magnitude stronger (in water) as an acid than NH_4^+, with a pKa of 9.2 (in water). Likewise, NH_2^- is about 15 orders of magnitude stronger as a base than HO-. The increased strength of the dominant base in ammonia, as well as the corresponding enhanced aggressivity of ammonia as a nucleophile, implies that ammonia would not support the metabolic chemistry found in terrain life. Terrain life exploits compounds containing the C=O carbonyl unit. In ammonia, carbonyl compounds are (at the very least) converted to compounds containing the corresponding C=N unit. Nevertheless, hypothetical reactions that exploit a C=N unit in ammonia can be proposed in analogy to the metabolic biochemistry that exploits the C=O unit in terrain metabolism in water (Haldane, 1954).

Given this adjustment, metabolism in liquid ammonia is easily conceivable. But most important: the intrinsic property of melanin to split and reform the water molecule is highly specific; the substrate must be water. If melanin is dissolved or combined with CN_4 melanin itself could be destroyed and the reaction that gives origin to the Life cannot take place, therefore the first requirement of any chemical reaction, energy; will be not available and therefore in the next step of the set of reactions nothing will happens.

Ammonia [NH_3] is only a polar solvent that might serve as an alternative to water but not as substrates in which melanin could take away the extraordinary capacity to split and reform the water molecule.

Formamide is another potential biosolvent.

Formamide is formed by the reaction of hydrogen cyanide with water; both are abundant in the cosmos. Like water, formamide has a large dipole moment and is an excellent solvent for almost anything that dissolves in water, including polyelectrolytes as DNA and RNA.

In formamide, many species that are thermodynamically unstable in water with respect to hydrolysis are stable. Indeed, some are spontaneously synthesized. This includes ATP (from ADP and inorganic phosphate), nucleosides (from ribose borates and nucleobases), peptides (from amino acids), and others (Schoffstall, 1976; Schoffstall et al., 1982; Schoffstall and Liang, 1985). Formamide is itself hydrolyzed by water, meaning that it persists only in a relatively dry environment, such as a desert.

But the intrinsic property of melanin to split and reform the water molecule cannot be expressed with formamide, only with water.

We might also consider non polar solvents. Hydrocarbons, ranging from the smallest (methane) to higher homologs (ethane, propane, butane, etc.) are abundant in the solar system. Methane, ethane, propane, butane, pentane, and hexane have boiling points of ca. 109, 184, 231, 273, 309, and 349 K, respectively, at standard terrain pressure.

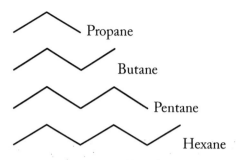

But why not use the hydrocarbons that are naturally liquid on Titan, the largest moon of Saturn, as a solvent for life directly? Broad empirical experience shows that organic reactivity in hydrocarbon solvents is no less versatile than in water. Indeed, many terrain enzymes are believed to catalyze reactions by having an active site that is not like water. Furthermore, hydrogen bonding is difficult to use in the assembly of supramolecular structures in water. The reactivity water means that it destroys hydrolytically unstable organic species. Hydrocarbons seems as would be able to guide reactivity more easily than life in water. However the problem is the same: Melanin splits and reforms only the water molecule. The energy released by melanin is indispensable as starting point in a self-sustainable chemical system that eventually goes into some kind of Darwinian evolution.

Other liquids are still more abundant. For example, the most abundant compound in the solar system is dihydrogen, the principal component (86%) of the upper regions of the gas giants, Jupiter, Saturn, Uranus, and Neptune. Throughout most of the volume of gas giants where molecular dihydrogen is stable, it is a supercritical fluid. For most of the volume, however, the temperature is too high for stable carbon–carbon covalent bonding. Dihydrogen becomes a supercritical fluid with a critical temperature of 33.3 K and a critical pressure of 12.8 atm. Above the critical point, a substance is a supercritical fluid, neither liquid nor gas. The properties of supercritical fluids are generally different than those of the regular fluids. In instance, supercritical water formed at a critical temperature of 647 K and a critical pressure of 112 atm is relatively non polar and acidic.

Furthermore, the properties of a supercritical fluid, such as its density and viscosity, change with changing pressure and temperature, and especially dramatically as the critical point is approached. Supercritical water below the earth surface leads to the formation of many of attractive crystals that are used in jewelry.

Little is known about the behavior of organic molecules in supercritical di-hydrogen as a solvent.

This discussion summarizes only some of the issues that might be addressed as we attempt to combine the constraints of physical and chemical laws, the consequent behavior of solvents (like water), and the structure of biological molecules. Synthesis is an experimental approach to better understanding of this interaction, and especially for how underlying physical law has constrained biology at the molecular level. It also provides intellectual processes that manage the natural tendency for scientists to become advocates for their individual prejudices.

CO_2 and Human Photosynthesis

- Without a steady supply of Energy, cells die within seconds.

- Thousands of different chemical reactions occur constantly within cells. The term **metabolism** (from the Greek: change) refers to the sum of all of the chemical reactions. These reactions are crucial to providing cells with the primary building blocks (amino acids, sugars, fatty acids) and their posterior transformation in more complex molecules.

- Every single chemical reaction of thousands of different that occur constantly within all cells has a very first common requirement: Chemical Free Energy.

Photosynthesis in Plants

As part of the carbon cycle known as photosynthesis, plants, algae, and cyanobacteria absorb carbon dioxide, light, and water to produce carbohydrate or carbon chains to make the biomass, energy is taken from water, and oxygen as a waste product.

Photosynthesis is the process by which organisms that contain the pigment chlorophyll convert light energy into chemical free energy which can be used to organize the biomass, that is based in carbon chains of different lengths, combinations, rotation; ETC. Energy cannot be stored in the molecular bonds of organic molecules (e.g., sugars). Photosynthesis powers almost all trophic chains and food webs on the Earth.

The net process of photosynthesis is described by the following equation:

$$6CO_2 + 6H_2O + \text{Light Energy} = C_6H_{12}O_6 + 6O_2$$

This equation simply means that carbon dioxide from the air and water combine in the presence of sunlight to form sugars; oxygen is released as a by-product of this reaction.

During the process of photosynthesis, light penetrates the cell and passes into the chloroplast. The light energy is intercepted by chlorophyll molecules on the granal stacks. Significant part of the light energy is converted to chemical free energy.

During this process, a phosphate is added to a molecule to cause the formation of ATP. The third phosphate chemical bond contains the new chemical energy. The ATP then provides energy to some of the other photosynthetic reactions that are causing the conversion of CO_2 into sugars.

While the above reactions are proceeding CO_2 is diffusing into the chloroplast. In the presence of the enzyme Rubisco, one molecule of CO_2 is combined with one molecule of RuBP, and the first product of this reaction is two molecules of PGA.

The PGA then participates in a cycle of reactions that result in the production of the sugars and in the regeneration of RuBP. The RuBP is then available to accept another molecule of CO_2 and to make more PGA.

Energy Incident on a Leaf

Photosynthesis is not a very efficient process but is sustainable at most. Of the sunlight reaching the surface of a leaf, approximately:

- 75% is refracted
- 15% is reflected
- 5% is transmitted through the leaf
- 4% is converted to heat energy
- 1% is used in photosynthesis

How do we know the O_2 is derived from H_2O during photosynthesis?

The oxygen product of photosynthesis could originate from either the CO_2 or the H_2O starting compounds. To determine which of these original compounds contributed to the O_2 end product, an isotopic tracer experiment was performed using ^{18}O:

- ^{18}O is a heavy isotope of oxygen
- $H_2^{18}O + CO_2$ yields $^{18}O_2$
- $H_2O + C^{18}O_2$ yields O_2

Therefore, the O_2 end product must originate from water and not from the carbon dioxide.

How do we know what the first products of photosynthesis are?

Another isotopic tracer experiment:

^{14}C is a radioactive isotope of carbon. $^{14}CO_2$ is exposed for a brief period to a green plant that is conducting a photosynthesis in the presence of sunlight. Immediately after exposure to $^{14}CO_2$, the plant's photosynthetic tissue is killed by immersing it in boiling alcohol, and all of the biochemical reactions cease. The chemical compounds in the dead tissue are all extracted and studied to determine which of them possesses the ^{14}C. Following the briefest exposure to $^{14}CO_2$, the only chemical compound that possessed ^{14}C was PGA (phosphoglyceric acid, a three carbon molecule). Following longer periods of exposure, much of the ^{14}C was found in a variety of compounds including glucose. By varying the length of the exposure period it was possible to identify the sequence of the reactions leading from PGA to glucose.

Metabolism

We have seen how plants convert sunlight (photonic energy) into sugars (carbon chains essentially). Now we need to understand how cells can use the products of photosynthesis (carbon chains) to organize biomass and not to obtain energy. There are several possible metabolic pathways by which cells can use sugars as building blocks of biomass; the phrase: energy stored in chemical bonds is a deep mistake, for instance: energy cannot be stored.

- o   Glycolysis
- o   Fermentation
- o   Cellular respiration

Glycolysis:

Glycolysis can occur in either the absence or the presence of oxygen. During glycolysis, glucose is broken down to pyruvic acid, an important metabolic intermediate; yielding 2 ATP moles. If Glycolysis occurs in the cytoplasm of cells, not in organelles, the energy for organelles, from where? Glycolysis occurs in all kinds of living organisms; because sugars are the perfect brick to build up the biomass. Prokaryote cells use glycolysis and the first living cells most likely used glycolysis.

Fermentation:

During fermentation, the pyruvic acid produced during glycolysis is converted to either ethanol or lactic acid. This continued use of pyruvic acid during fermentation permits glycolysis to continue with its associated production of ATP, however, ATP is not a source of energy. If glucose were a source of energy, then diabetic patients must be able to climb walls.

Cellular Respiration:

Respiration is the general process by which organisms oxidize organic molecules (e.g., sugars) and derive energy (ATP) from the molecular bonds that are broken. This is the ancient interpretation. However, the discovery of the human photosynthesis shows that ATP is a metabolic pathway whose main aim is the drive of phosphate groups; by other side, when ATP is converted to ADP, energy is absorbed.

Glucose (a sugar):

C $_6$H$_{12}$O$_6$

, and is described by the equation:

$$C_6H_{12}O_6+6O_2 \text{----------}> 6CO_2+6H_2O+36ATP$$

Simply stated, this equation means that oxygen combines with sugars to break molecular bonds, releasing molecules which rearrangement will form organic compounds.

The energy released is so little and the final products of the reaction are carbon dioxide and water.

In eukaryotic cells, cellular respiration begins with the products of glycolysis being transported into the mitochondria. A series of metabolic pathways (the Krebs cycle and others) in the mitochondria result in the further breaking of chemical bonds and the liberation of ATP. CO_2 and H_2O are end products of these reactions. The entirely theoretical maximum yield of cellular respiration is 36 ATP per molecule of glucose metabolized. Notice: is entirely theoretical. By other side, NADH is also used as reducing agent for many cellular reactions.

Note that photosynthesis is a reduction-oxidation reaction, just like respiration). In respiration energy is released from sugars when electrons associated with hydrogen are transported to oxygen (the electron acceptor), and water is formed as a byproduct. The mitochondria use the energy released by photosynthesis in the oxidation of sugars in order to synthesize ATP. In plants´photosynthesis, the electron flow is reversed, the water is split (not formed), and the electrons are transferred from the water to CO_2 and in the process the energy is used to reduce the CO_2 into sugar. In respiration the energy yield is 686 kcal per mole of glucose oxidized to CO_2, while photosynthesis requires 686 kcal of energy to boost the electrons from the water to their high-energy perches in the reduced sugar -- light provides this energy.

Differences between Prokaryote and Eukaryote

What makes a eukaryote so different from a prokaryote is its variety of organelles; many of which are believed to once have been prokaryotes in their own right. One of the most distinctive organelles is the mitochondria. Mitochondria, or mitochondrion, are the energy producing organelles of the eukaryotic cell[5]. Mitochondria is the only organelle that posses a double membrane leading many scientists to believe that the mitochondria used to be prokaryotic cells that where swallowed up by a larger cell. This double membrane serves to increase surface area on the inside of the cell. This convoluted internal membrane is called the cristae of the mitochondria. This is coiled up so such a large extent that it makes up one third of a eukaryotic cell's total membrane. This membrane projects into an inner space called the matrix. The matrix houses the mitochondrial DNA and ribosomes. The

---

[5] Where is the powerhouse of Mitochondria itself?

existence of mitochondrial DNA further supports the theory that the mitochondria used to be separate cells. The matrix is also home to highly concentrated mixtures of enzymes that assist in breaking down carbohydrates[6]. This in turn produces ATP, which is so far wrongly believed that is the main power source of the cell[7]. This process is called cellular respiration. Respiration is one of the key ways a cell gains useful energy to fuel cellular changes._The reactions involved in respiration are catabolic reactions that involve the redox reaction (oxidation of one molecule and the reduction of another). Presently organisms that use oxygen as a final electron acceptor in respiration are described as aerobic, while those that do not are referred to as anaerobic.

The energy released in respiration is used to synthesize ATP[8] to store this energy. The energy stored in ATP can then be used to drive processes requiring energy, including biosynthesis, locomotion or transportation of molecules across cell_membranes[9].

---

[6] And assist too to the breakdown of ATP to ADP. When ADP is build up to ATP energy is released, when ATP is broken down to ADP, then energy is absorbed.

[7] Human Photosynthesis breaks the paradigm: neither Glucose nor ATP is source of energy.

[8] It is not possible to store energy. ATP is an important step in the interaction between energy and biomass, but the main role of ATP is the modulation of metabolic intermediate.

[9] ATP, its precursors and derivatives have a key role in the Biosynthesis of numerous biological molecules but mainly and in certain cases only as source of building blocks, not as source of energy.

Most of the ATP produced by aerobic cellular respiration is made by oxidative phosphorylation. This works by the energy released in the consumption of pyruvate being used to create a chemiosmotic potential by pumping protons across a membrane. This potential is then used to drive ATP synthase and produce ATP from ADP and a phosphate group. Biology textbooks often state that 38 ATP molecules can be made per oxidized glucose molecule during cellular respiration (2 from glycolysis, 2 from the Krebs cycle, and about 34 from the electron transport system). However, this maximum yield is never quite reached due to losses (leaky membranes) as well as the cost of moving pyruvate and ADP into the mitochondrial matrix and current estimates range around 29 to 30 ATP per glucose.

Therefore the scarce ATP production by means of glucose oxidation is diminished even more when other factors are taken into account.

Aquatic environment

Phytoplankton and algae in aquatic environments account for almost 50% of the carbon dioxide (CO_2) fixed by photosynthesis on the Earth. Because the diffusion of CO_2 is almost 10,000 times slower in water than in air, almost all of these organisms have a mechanism that concentrates CO_2 from the environment for photosynthesis. This CO_2 concentrating mechanism (CCM) is essential for photosynthesis at atmospheric levels of CO_2 and for the survival of algae. In photosynthetic cyanobacteria, the CCM includes membrane proteins that transport bicarbonate (HCO_3-) from the medium into the cell and a carbonic anhydrase. The carbonic anhydrase is needed to convert the accumulated HCO_3- to CO_2 for Rubisco, the enzyme that fixes CO_2. The CCM in eukaryotic algae is less well understood but is known to involve transport proteins and multiple carbonic anhydrase.

Carbon dioxide (CO_2) normally comprises about 385 parts per million (ppm) in the atmosphere. The partial pressure of CO_2 in the blood is normally about 40mm of mercury (mmHg). A high carbon dioxide level (hypercapnia) is generally defined as a CO_2 pressure of 45 mmHg and 75 mmHg is considered to be severe hypercapnia. Mild cases of hypercapnia do not typically require specific treatment other than breathing normally carbonated air, but severe hypercapnia may require prompt medical attention.

Carbon dioxide produced by body tissues diffuses into the interstitial fluid and into the plasma. Less than 10% remains in the plasma as dissolved CO_2. The rest (70%) diffuses into red blood cells, where some (20%) is picked up and transported by hemoglobin. Most of the CO_2 reacts with H20 in the red blood cells to form carbonic acid. Red blood cells contain the enzyme carbonic anhydrase, which catalyzes this reaction. Carbonic acid dissociates into a bicarbonate ion and hydrogen ion (H+). Hemoglobin (a plasma protein) binds most of the H+, preventing them from acidifying the blood. The reversibility of the carbonic acid- bicarbonate conversion also helps buffer the blood, releasing or removing H+ depending on the pH. Chlorine goes into the red blood cells when bicarbonate comes out. This is referred to as the chloride shift.

Carbon dioxide produced by the human body is not a toxin. Carbon dioxide levels in the blood are not dangerous until they reach high levels. This gaseous metabolic byproduct typically exits the body quickly. Once it dissolves into the blood stream, it becomes bicarbonate and is flushed out by the kidneys or carried to the lungs, transformed back into carbon dioxide and exhaled. The same elimination process expels inhaled carbon dioxide.

This is the traditional simplified reaction:

$$C6H12O6 + 6O2 \rightarrow 6CO2 + 6H2O + energy$$

However, from the point of view of human photosynthesis, the equation is this:

$$C6H12O6 + 6O2 \rightarrow 6CO2 + 6H2O + biomass\ (not\ energy)$$

About 70 percent of the air a person inhales fills the millions of alveoli in the lungs. Each alveolus is surrounded by a basketlike mesh of blood capillaries. The wall separating the inhaled air from the blood is only 0.5 micrometer thick—only one-fifteenth the diameter of a single red blood cell—so it presents very little barrier to gas diffusion between the air and blood.

Oxygen has a concentration (partial pressure) of 104 mmHg in the alveolar air and 40 mmHg in the arriving capillary blood. Thus, it diffuses down its concentration **gradient** from the air, through the alveolar wall, into the blood. About 98.5 percent of this oxygen binds to the pigment **hemoglobin** in the red blood cells, and the other 1.5 percent dissolves in the blood plasma.

Carbon dioxide (CO_2), has a partial pressure of 46 mmHg in the arriving blood and 40 mmHg in the alveolar air, so its concentration gradient dictates that it diffuses the other way, from blood to air, and is then exhaled. About 70 percent of this CO_2 comes from the breakdown of carbonic acid in the blood; 23 percent from CO_2 bound to hemoglobin, albumin, and other blood **proteins**; and 7 percent from gas dissolved in the blood plasma.

Blood typically contains 95 mmHg O_2 upon arrival at the systemic capillaries and 40 mmHg O_2 upon leaving. Conversely, the blood has 40 mmHg of CO_2 on arrival at the systemic capillaries and typically 46 mmHg CO_2 when it leaves. The blood does not, however, unload the same amount of O_2 to all tissues or pick up the same amount of CO_2. The more active a tissue is, the warmer it is, the lower its O_2 level is, and the lower its **pH** is (because it generates more CO_2 and CO_2 reduces the pH of body fluids). Heat, low O_2, low pH, and other factors enhance O_2 unloading and CO_2 loading, so tissues that need the most oxygen and waste removal get more than less active tissues do. The biochemistry of hemoglobin is mainly responsible for this elegant adjustment of gas exchange to the individual needs of different tissues.

Chapter 26

Human Photosynthesis, the Ultimate Answer to the Long Term Mystery of Kleiber's Law or E=M ¾

Authors:

Arturo Solís-Herrera, MD., PhD.*, Ruth I. Solís-Arias, MD., Paola E. Solís-Arias, MD., Martha P. Solís-Arias, MD, María del Carmen Arias Esparza, MD.

*Human Photosynthesis Study Center.

Av. Aguascalientes Norte 607, Pulgas Pandas Sur, CP 20230; Aguascalientes, México. comagua2000@yahoo.com

Abstract:

Kleiber's Law or $E = M^{\frac{3}{4}}$ is a mathematical expression known since 1932 that outlines the relationship between mass (biomass) and the use of energy. It is compelling because it resumes the longstanding observation that big animals use energy more efficiently than little ones. An elephant is 200,000 heavier than a mouse, but uses only about 10,000 time more energy; thus a cat, having a mass 100 times that of a mouse has a metabolism roughly 31 times greater than that of a mouse. In other words, the bigger you are, you actually need less energy per gram of tissue to stay alive. That is an amazing fact[53]. Many facts pertaining to animal size call for rational explanations.

It is fascinating that the relationship between mass and energy use for any living thing is governed strictly by a mathematical formula universal across all of life. It operates in the tiniest of bacteria to the biggest of whales and sequoia tress. Even though $E = M^{3/4}$ was discovered back in the 1930's, no one has been able to explain it satisfactorily. Nevertheless, our discovery of Human Photosynthesis or the hitherto unknown capability of the mammal eukaryotic cell to split, break or dissociate the water molecule through melanin, first seen in human retina and then in all eukaryotic cells, finally unravels this mystery: the bigger you are the more surfaces you have to absorb electromagnetic radiations, or bigger you are better antenna you are..

Key words: Kleiber's Law, Melanin, Photosynthesis, energy, water dissociation.

Organism Mass		
Mycoplasma	<0.1 pg	$<10^{-13}$ g
Bacterium	0.1 ng	10^{-10} g
Tetrahymena	0.1 µg	10^{-7} g
Amoeba	0.1 mg	10^{-4} g
Bee	100 mg	10^{-1} g
Hamster	100 g	10^{2} g
Human	100 kg	10^{5} g
Blue whale	>100 tons	$>10^{8}$ g

Table 1. Each step represents a 1000-fold difference in mass. Modified from Schmidt-Nielsen, Knut. Scaling, Why is Animal Size so Important. Cambridge University Press. 1984. ISBN 0 521 26657 2

Background:

When we try to find the rules that govern animal functions, we tend to think in terms of chemistry. We think of water, salts, proteins, enzymes, oxygen, energy and so on, a whole word of chemistry. However, the physical laws are equally important; they determine rates of diffusion and heat transfer, transfer of force and momentum, the strength of structures, the dynamics of locomotion and so on. Physical laws provide opportunities and possibilities, impose constraints and set limits to what is physically possible[54].

The world we live in is governed by the laws of chemistry and physics, and animals must live within the bounds set by those laws. The body size has profound consequences for structure and function and the size of an organism is of crucial importance to the question of how it manages to survive.

Life is considered the most complex and diverse physical phenomenon in the Universe, manifesting an extraordinary diversity of form and function over an enormous scale from the largest animals and plants to the smallest microbes and sub-cellular units. Despite this many of its most fundamental and complex phenomena scale with size in a surprisingly simple fashion[55]. The metabolic rate scales as the ¾ power of mass over 27 orders of magnitude, from molecular and intracellular levels up to the largest organisms, are good examples of this. The smallest shrew, when fully grown, is only one-tenth the size of a mouse, or one-millionth the size of an elephant; however, the elephant don´t eat a million times more than the shrew. In example, the Etruscan Shrew has a weight of 2 g, and requires 1.428 g daily of food, in contrast, an adult elephant consumes 140 to 170 kg of food a day, only 100 000 times more than the shrew.

Organisms themselves span a mass range of over 21 orders of magnitude, ranging from the smallest microbes (10^{-13} g) to the largest mammals and plants (10^{8} g). Despite this amazing diversity and complexity, many of the most fundamental biological processes manifest an extraordinary simplicity when viewed as a function of size, regardless of the class or taxonomic group being considered.

Scaling with size typically follows a simple power law behavior of the form:

$$Y = Y_o \, M_b^{\,b}$$

Where Y is some observable biological quantity, Y_0 is normalization constant, and M_b is the mass of the organism. An additional simplification is that the exponent b, takes on a limited set of values, which are typically simple multiples of ¼. Among the many variables that obey these simple quarter-power allometric scaling laws are nearly all biological rates, times, and dimensions; they include metabolic rate ($b\approx3/4$), lifespan ($b\approx1/4$), and heart rate ($b\approx1/4$), DNA nucleotide substitution rate ($b\approx1/4$), lengths of aortas and heights of trees ($b\approx1/4$), radii of aortas and tree trunks ($b\approx3/8$), cerebral gray matter ($b\approx5/4$), densities of mitochondria, chloroplast and ribosome ($b\approx-1/4$), and concentrations of ribosomal RNA and metabolic enzymes ($b\approx-1/4$).

The best-known of these scaling laws is for basal metabolic rate, which was first shown in 1932 by Max Kleiber, author of the thesis "The Energy Concept in the Science of Nutrition", who came to the conclusion that the ¾ power of body weight was the most reliable basis for predicting the basal metabolic rate (BMR) of animals and for comparing nutrient requirements among animals of different size.

Subsequent researchers showed that the whole-organism metabolic rates also scale as $M_b^{3/4}$ or Kleiber's law, in nearly all organisms, including animals (endotherms and ectotherms, vertebrates and invertebrates), plants, and unicellular microbes. This simple ¾ power scaling has now been observed at intracellular levels from isolated mammalian cells down through mitochondria to the oxidize molecules of the respiratory complex, thereby covering fully 27 orders of magnitude[56]. The enormous size differences (Table 1) among living organism are no easily conceptualized, i.e. the total difference between the smallest and largest organism –blue whale- is 10^{21}, an hypothetical giant organism larger than the blue whale by the same ratio, 10^{21} ; would be 100 times the volume of the earth. Our universe, with its incomprehensible magnitude have a total mass of some 10^{80} gram.

Young (i.e., small) organisms respire more per unit of weight than old (large) ones of the same species, once the organism is out of the amniotic fluid. Inside the uterus the fetus breathing is slow, e.g. 20-30 per hour versus 20 to 40 per minute after birth. Traditionally, this has been explained by the overhead costs of growth, furthermore, some authors state that small adults of one species respire more per unit of weight than large adults of another species because a larger fraction of their body mass consists of structure rather than reserve, and structural mass involves maintenance costs where reserve mass does not. However when we analyze Kleiber's law from the point of view of Human Photosynthesis, we see this is not the issue.

The exponent in Kleiber's law, a power law where a mathematical relationship exists between two quantities, has been a matter of dispute for many decades. It is still contested by a diminishing number as being ⅔ rather than the more widely accepted ¾. Given the fact that this law is concerned with the capture, use, and loss of energy by a biological system, the system's metabolic rate was at first taken to be ⅔, because energy was thought of mostly in terms of heat energy, thus metabolic rate was expressed in energy per unit time, specifically calories per second.

This misconception was based on the wrong belief that energy and biomass evolve from the food we ingest. Two thirds therefore expressed the ratio between surface area (length^2) and volume (length^3) of a sphere, with the volume of the sphere increasing faster than the surface area, with increases in radius. This surface area-to-volume ratio gave the metabolic rate of a particular organism as proportional to its mass raised to the power of $^2/_3$. This was purportedly the reason large creatures lived longer than smaller ones - that is, it was thought that as they got bigger they lost less energy per unit volume through the surface, as radiated heat. Regardless of the exponent, ⅔ or ¾, what has not yet been considered is the fact that a bigger surface, sphere, or mass, gives an organism better capability to absorb the electromagnetic radiations of diverse wavelengths, an unexpected and yet existing source of energy in chlorophyll-lacking living organisms.

The concept of metabolic rate itself was poorly defined and difficult to measure. It seemed to concern more than just the rate of heat generation and loss. Prevailing understanding of an organism's metabolic/respiratory chain was based entirely on blood-flow considerations. And yet, the equation has been shown to be relevant over 27 orders of magnitude, extending from bacteria, which do not have hearts, to whales or forests. Therefore, to assume that this is due to how resources are distributed through hierarchical branching networks it is not a good explanation, given that the role of fractal capillary branching is not demonstrated as fundamental to the exponent ¾.

Blood flow and metabolic efficiency (ME)

The theoretical models are also relevant to things without blood flow, like bacteria and coral. Attempts to understand the metabolic rate of a multi-cellular organism (field metabolic rate, which includes the activity of the organism) are expressed in terms of the product between average basal metabolic rate, and number of cells. This, plus capillary terminal size invariance, leaves the equation open to the criticism that it cannot possibly account for spikes in metabolic rate needed for motor activity. Too much blood would be required.

In plants the exponent of mass is close to 1^{57}. Mathematically, this is not possible since the implication is that ME is greater than 100% in the case of plants and between 89 and 100 % in mammals. Efficiencies like these are not found in Nature.

The term Calories is a measure of heat energy. Undesirably this leads to the idea that thermo-genesis is part of metabolism, a mistake in spite of the fact that it was Kleiber's original treatment, and it also does not consider that metabolism is about chemical energy, not heat energy. The picture is further confused when the idea of respiratory metabolism is introduced refining and limiting the definition of metabolism, making oxygen consumption and synthesis of ATP its ultimate factors. When considering metabolic rates of cells in vitro, data from studies of oxygen consumption suggest that the exponent is not only far less than ¾, but even becomes negative for things less than one gram in size. Furthermore, this model excludes glyco-genesis from metabolic consideration since glyco-genesis is not included in the respiratory chain, and is itself a reduction reaction not strictly dependent upon the proximity of certain molecules and atoms delivered by capillaries and vibrating from Brownian motion. Energy is required for glycogenesis, and blood does not deliver energy, just the ingredients for endergonic reactions. The energy comes from redox coupling, what ME is all about.

ME amend these problems, and describe the metabolic rate in watts. Metabolic rate becomes the rate at which a biomass recharges so that its degeneration is prevented, and its organization is perpetuated. ME is a ratio of the rate of reduction reactions necessary for the maintenance, growth, replication and behavior of the biomass, to the rate of availability of energy captured and expended by that biomass. ME is a statement of redox coupling efficiency. ME consequently excludes thermogenesis as part of metabolism. A graph of Kleiber that includes ME, with ME as the X axis, metabolic rate as the Y axis, and a different curve for each mass, reveals a picture of the relation of biomass to metabolic rate that suggests all of evolution took place at less than 45% ME. The organism determines ME, and that ME is the same for it and for its cells. If we put a limit in the considerations of metabolism to strictly electrochemistry and then, if metabolic rate (MR) is taken as the recharge rate in watts of an electrolytic biomass, then MR is directly related to the longevity of that biomass.

The resultant equation suggests that for living things operating at over 25% ME, the organism lives longer than its cells.

What this indicates is that the organism has a source of new cells that are not initially a part of it, undifferentiated stem cells for example, especially in creatures over one gram mass. Aging appears to be the result of antagonism between BMR (basal metabolic rate or each cell) and FMR (field metabolic rate or the entire organism), given fluctuations in ME. These fluctuations are driven predominantly by alterations in the denominator of the ratio ME, where that denominator represents the availability of food sources.

Kleiber's law, as originally formulated, was based upon the idea that metabolic energy was entirely related to measurements of heat generation and loss. On the other hand, trying to explain part of Kleiber's law with blood flow is doomed by the invariant size of capillaries, which is the same in leaves and mammals.

Resting Energy Expenditure

The relationship between resting energy expenditure (REE) (kJ/d) and body mass (M)(kg) is a cornerstone in the study of energy physiology. Kleiber formulated the now classic equation: $REE=293M^{0.75}$. The biological processes underlying Kleiber's law have been a topic of long standing interest and speculation[58].

All living mammals expend energy for the maintenance of the resting energy expenditure (REE), the thermic effect of feeding, and for physical activity. REE is usually the largest portion of total energy expenditure.

Kleiber´s law is one of the most important and best-known laws in bioenergetics[59]. The consistency of the REE/M relation over so wide a range of body sizes end species suggests some unique biological characters inherent within this relation. In the following years, there was –and still was- considerable discussion devoted to explaining Kleiber law. Although a number of hypotheses have been proposed, there is not yet a fully agreed upon mechanistic understanding of 0.75 exponent observed across mature mammals[60],[61],[62].

Life and energy

Living organisms are capable of generating order from disordered matter. They are also capable of replicating themselves with great fidelity. In both cases there are two important requirements, the first one: energy, the second: building blocks or carbon chains. The workings of living beings are intricate; they rely on molecular mechanisms that are quite delicate, very fine-tuned yet robust to the vicissitudes of life and to their environment. We need to center on tracking the flow of available energy in living systems. This common thread illuminates an underlying simplicity[63].

All living systems are out of thermodynamic equilibrium with their surroundings. A necessary condition for producing, as well as for maintaining, a dynamic system out of equilibrium is to be able to access available external energy. Life requires a careful "balanced imbalance."

Available energy in biology comes in several forms: photon energy, chemical energy, electrical energy, and mechanical strain are the common ones. This available energy, also called high-grade energy, has to be conserved or utilized and not completely lost as heat to the surroundings, for metabolism to function in living systems. In fact, in most metabolic processes, available energy is converted from one form to another, although never with 100% efficiency.

Living matter, on a large scale, has a well-defined temperature, pressure, density and chemical composition. In short, it is in a thermodynamic state or steady state. Although there are statistical fluctuations, they are negligibly small. In contrast, isolated single molecules have to be described by the laws of classical or quantum mechanics. In a volume as big as a mitochondrion at pH 8 there is only a single proton present on the average. The transfer of a single proton across the membrane, therefore, changes pH appreciably. This highlights the fact that all thermodynamic properties of such small systems undergo appreciable fluctuations in space and time requiring constantly energy and energy dotted with certain characteristics as delimited range of fluctuation, chemical in nature, enough all time for the hundreds or thousands of reactions that occur incessantly into the cell or organism.

Thermodynamic states of matter are fully described by a small number of thermodynamic variables: the total volume, temperature, pressure, density, available energy, chemical composition and concentrations of the chemical components, and in mammals, human being included: amount of light or photonic energy. More important: when energy is degraded into heat, it is not recovered by living systems[64].

Melanin, a unique compound

Melanin appears to be a candidate for primordial catalysis in nature prior to the evolution of proteins. The reluctance to accept melanin as biologically active substance was based on its stability[65]. The next brief paragraphs about enzymes and catalysis show examples of activities that melanin exhibits, however melanin is much more than that.

The detailed evolution of a reaction has a complex dependence on its environment. Here it will be referred to simply as the reaction path. Within a given environment, whatever the path, ΔG (the change in free energy) depends only on the reactants and the products. Therefore ΔG is treated as discreet variable; otherwise if it were treated as a continuous variable, the result would exceed our abstraction capability.

The speed of a chemical reaction depends on its detailed path. In ordinary reactions, a great simplification results from postulating a transition state that constitutes the bottleneck for the reaction. In this approximation, the transition state is produced by local, microscopic fluctuations within the thermodynamic state. However these microscopic fluctuations depend of the energy as first requirement and of the result of it over the present matter in second term.

Given a set of reactants, usually many reaction paths are possible. Among those, the one with the fastest rate predominates. A crucial observation is that the reaction rate is determined by the "activation" free energy of the transition state, $\Delta G\dagger$, and not by the final ΔG of the reaction. In particular, some reaction paths can generate a large amount of heat, *i.e.* have a very large negative ΔG, while some other reaction path with a much smaller negative ΔG and different reaction products can be a faster reaction. That is the case when the transition state energy $\Delta G\dagger$ of the second reaction is lower, and/or its crossing probability is larger than those for the first one. A similar argument can be given for generalized reactions if a bottleneck for the reaction can be identified.

Catalysis in Brief

Catalysis, defined in the broadest possible way, including any change in the reaction path caused by the catalyst, encompassing the change of speed of a given reaction, and including changes in the products, has as a prerequisite for the catalyst to be able to influence a chemical reaction that a high-grade energy be exchanged among the catalyst and the reactants; or, expressed even more precisely, the catalyst has to interact with the reactants in order to influence the reaction. Thus quantitative predictions can be obtained only by solving the microscopic equations of classical or quantum mechanics respectively.

Catalysts provide a local environment to the reaction. As long as the reactants are in their macroscopic, thermodynamic states this is a valid and useful description. The reactants are then in local equilibrium before the reaction: they have a local G, a local pH and some chemical groups have locally varying pK. Most catalysts provide a specific environment for a specific reaction, just like melanin does for water dissociation and reformation. Similarly, the products are in local equilibrium after the reaction with a different local G, etc. At this intermediate stage ΔG of the reaction depends on the catalyst. This apparent contradiction is resolved if ΔG of binding the reactants and of unbinding the products is properly added. The resulting total ΔG is independent of the catalyst, as it should be, and melanin is not an exception.

The importance of catalysis for biology is that catalysts can influence the pathway of the reaction in order to accomplish a biological task. For example, catalysts are able to change the path of a reaction in such a way that available energy is sequestered during the reaction and, possibly, channeled into new products. By selecting a reaction path, catalysts can change the fraction of the available energy, G, in each of the steps, while keeping the sum of ΔG constant for all steps in the sequence. In such reactions, even when a catalyst does not change the dissipated fraction of G, it can alter the path of the reaction in such a way that it temporarily holds some of the available energy during one part of the reaction, releasing it in a later part of the reaction. The fundamental mechanism of coupled reactions, ubiquitous in metabolism, is the exchange of high-grade energy between the exoergic and the endoergic components. We can state quite generally that the channeling of a reaction along a pathway of low heat loss, by enzymes, is the most important way high-grade energy is utilized in biology. We could consider melanin as a kind of enzyme whose relevance has not yet been taken into account in the sequence of the life, or in other words in any biochemical reaction.

Emergent laws

It is accepted wisdom that properties of living systems are based on well-established laws of physics and chemistry[66],[67],[68]. Nevertheless, the rules that govern complex systems in general and biological systems in particular cannot be directly deduced from those laws. In fact, they have to be described and understood on several distinct layers or levels. On each level a set of rules, appropriate to that level, has to be established. These properties have been called emergent behavior. Emergent laws of biology must not contradict the laws of physics and chemistry but they have to be established in addition to them[69].

In the utilization of available energy by living systems, as the driving force of life, we can consider three levels: macroscopic, mesoscopic and microscopic. At the macroscopic level, thermodynamics is valid and fluctuations are negligible. The concepts of free energy and of reversible and irreversible processes are well defined. At the microscopic level, the level of quantum chemistry, the equations of motion are reversible and thermodynamic concepts lose meaning. The level being considered to explore the energetic of the basic metabolic processes is the mesoscopic level. It is intermediate between quantum chemistry and cell biology and can be used to describe energy utilization in a physically, chemically, and biologically meaningful way.

Bioenergetics and metabolism

Metabolism represents the set of chemical and physical changes that take place within an organism and enable its continued growth and functioning. Living organisms exchange material and energy with their surroundings; and, from a thermodynamic point of view, they are open systems. The engines of life that keep metabolism going are fueled by available energy, this is the most general expression of the principles of bioenergetics[70], and in plants and animals this energy comes from light. In the case of human beings there is a deep-seated belief that energy and building blocks come from food. Deciphering the capability of melanin to split the water molecule breaks the ground of bioenergetics. We could say that human beings are autotrophic organisms, as plants are.

Almost every reaction in metabolism is catalyzed. This is true even for generalized reactions in living systems, such as motion or protein folding. In brief and based on the utilization of available energy, we have:

Biological Catalysts – Enzymes

The fundamental mechanism of enzyme action is a transient, *reversible* exchange of high-grade energy between the enzyme and the reactants. In classic terms, the enzyme exchanges high-grade energy with a reactant reversibly; both the reactant and the catalyst are distorted when they bind. The change in free energy, ΔG, of the *unbound* reactants and the products in a specific reaction does *not* depend on the presence of the catalyst. Nevertheless, when the reactants and the products are bound to a catalyst, the local values of G of the reactants and the products depend strongly on the local environment provided by the enzyme. A reaction is speeded up when the bound reactant "approaches" the transition state.

The most widely recognized property of enzymes is that they can speed up spontaneous reactions by binding to the transition state more strongly than they bind to the reactants. That lowers the relative free energy of the transition state $\Delta G\dagger$. Note that we do not differentiate among the various ways this is accomplished: by lowering the transition state, unbinding the reactants and "pre-organizing" the enzyme itself.

It is especially important that key steps in bioenergetics involve the reversible exchange of high-grade energy between uphill and downhill chemical processes. Such "coupled reactions" are the backbone of biosynthesis.

At this point it is convenient to extend the definition of catalysis to a situation when the exo-ergic and the endoergic reactions are carried out consecutively. The exo-ergic reaction then "charges" the enzyme with high-grade energy and the endoergic reaction is driven by the energized enzyme. Only the net difference between the two ΔG's is dissipated as heat and, at the end of the cycle, the enzyme is reusable. In order to make each step spontaneous, ΔG has to be negative at each step separately.

The great flexibility enables selection of "suitable" catalysts by evolution. We note that many biochemical cycles, *e.g.* the citric acid cycle and the Calvin cycle are now included in our definition. Also, that different enzymes have different efficiencies for the same reactions[71],[72]. Heat generated during metabolism is not reconverted into available energy by living beings today. Some organisms seem to benefit from the generation of heat. There are many great examples: (i) Flowers have hot plates to evaporate attractants. (ii) Insects (e.g. the bumblebee) produce heat in futile cycles so that they can take off in the cold. (iii) Bears waking from hibernation heat up their body by un-coupling electron transport.

A framework of life

The emergent set of rules compatible with the basic laws of physics, even though they cannot be derived from these basic laws, explain the functioning of living matter and form the framework for catalytic energy utilization, providing the link between biology on one side and physics and chemistry on the other. These rules are the following:

(1) Dynamic systems require an input of external energy to move away from thermodynamic equilibrium and a continuous input of external energy to maintain a position away from equilibrium. We now know that in both, plants and animals, that continuous input comes from the sun.

(2) Dynamic, changing and evolving systems can stay out of equilibrium only if they do not degrade all of the available energy

into heat during chemical changes. In order to achieve that, the fastest chemical reactions have to proceed on paths that have a relatively small negative ΔG.

(3) A catalyst interacting with its reactants can exchange high-grade energy with the reactants and thereby alter the fastest path of the reaction. The sign of this transient energy exchange can be either positive or negative with respect to the catalyst, and when it is positive, the catalyst gets transiently energized. In the simplest coupled reactions, the energy is transmitted to another reactant and the net result is an activated molecule. The conserved high-grade energy from this step is then available to the organism for other uses. In more complex - and more common - coupled reactions, the temporarily stored high-grade energy in the catalyst becomes available to drive another reaction up-hill. This ability is a "built in" feature of catalysis, and follows from the transient exchange of high-grade energy between catalyst and reactants, as discussed here. Therefore, the energy released during water dissociation and reforming of the water molecule is used in many ways by the eukaryotic cell.

(4) Catalysts provide a very large range of possibilities for selecting and varying chemical reactions. An enzyme with less than perfect specificity allows evolution to work; actually, it is the prime vehicle of evolution.

As a consequence, in living systems, evolution selects catalysts that are best suited for a specific task. Replication and evolution are consequences of this relatively simple framework, and as such they are also emergent properties of biological systems. The new discovery needed to understand the workings of life, is centered in the fact that the main source of energy in the animal eukaryotic cell is the water, in contrast with the ancient belief that energy and building blocks come from food.

Melanin, the human chlorophyll.

Twenty-one years have passed since I began to study treatments for the three main causes of blindness; and our results were astonishing: Melanin has the amazing capability to dissociate the water molecule[73].

Briefly, the reaction could be described this way:

$$2H_2O \longleftrightarrow 2H_2 + O_2 + 4e^-$$

The human photosynthesis system is composed of Light/Melanin/Water, arranged in order of abundance in Nature. Melanin has the capability to harvest visible and invisible light, and with that energy melanin can break the water molecule[74]. By analogy with Plant photosynthesis, we call this capability "Human photosynthesis."

The difference with photosynthesis in plants is crucial:

$$2H_2O \longrightarrow 2H_2 + O_2$$

Chlorophyll just splits the water molecule, however is unable to reform the water molecule.

Therefore the reaction in our photo-system is bidirectional, where as in plants is unidirectional.

Chlorophyll only absorbs 400 (blue) and 700 (red) nanometers of wavelength. In contrast, melanin absorbs the full electromagnetic spectrum. Therefore our photosynthesis is clearly different and much more efficient than that of plants.

The unexpected finding that the eukaryotic cell has the capability to harvest photonic energy and use it to break the water molecule by means of a photo-system composed of Light/Melanin/Water, in order of abundance in nature, explains clearly the longstanding observation that animals with greater size require less energy to sustain life, and furthermore, see their lifespan increased significantly. We can easily understand that the organism is acting like a biological antenna; therefore, if the body in question has more surface area or a larger mass, has greater probabilities to catch energy in enough amounts to sustain life, or highly complex biochemical reactions interlaced between them, that always have energy as a first requirement. Therefore we think that the unsuspected capacity of eukaryotic cell to split the water molecule through the photo-system composed by Light-Melanin-Water could be the great unifying principle.

The absorption of light is a mass-dependant phenomenon as is shown in the positive lenses example (Figure 1 and 2), eukaryotic cell can be compared with a positive lens.

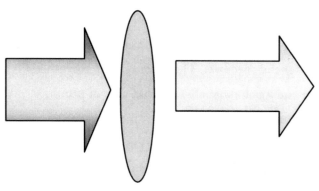

Figure 1. Incident light (a), positive lens (B) and emergent light (C) which is significantly absorbed.

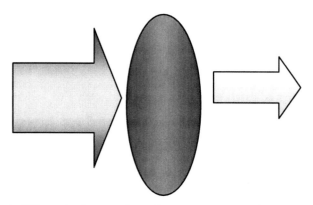

Figure 2. When the lens´ index refraction, curvature (to magnifies the image), or the lens body-mass is increased, the transmission of incident rays is then diminished due to lens absorption that is further raised.

COMMENT:

Evolution by natural selection is one of the few universal principles in biology, wrote West, Brown and Enquist in an article published in 4 June of 1999[75]. Evolution has shaped the structural and functional design of organisms in two important ways. First, they said, it has tended to maximize metabolic capacity, because metabolism produces the energy and materials required to sustain and reproduce life. And yet, the discovery of Human Photosynthesis, or the unexpected capability of melanin to split the water molecule, radically changes this widely accepted concept: Metabolism drives the interaction between energy and the building blocks that our body takes from the food we ingest, because metabolism is not an energy producing system, it solely represents the interplay between the building blocks obtained from food and our energy requirements; requirements that our organism takes from water.

West, Brown and Enquist also wrote that metabolic capacity has been maximized "by increasing surface areas where resources are exchanged with the environment." Under the principles of Human Photosynthesis this enunciate is correct: the larger the surface, the more exchange there is with the environment; that exchange is also applicable to light, or in a more descriptive way, to electromagnetic radiations, visible and invisible.

The second important achievement evolution has brought, as cited by West, Brown and Enquist, is the tendency to maximize internal efficiency by reducing distances over which materials are transported and hence the time required for transport. And a further consequence of evolution is the incredible diversity of body sizes, which range over 21 orders of magnitude, from 10^{-13} g (microbes) to 10^8g (whales). So, what are the consequences of a change in size?, are there upper and lower limits to the size of living organisms?

Mycoplasma or pleuropneumonialike organism (PPLO); is the smallest organism we know that is able to live and reproduce by itself in an artificial medium. It is so small, that, if the aqueous contents of the cell are at neutral pH, there will be, on the average, no more than two hydrogen ions inside the cell. Because the macromolecules that carry the metabolic and genetic functions are essential, and their size probably cannot be reduced, the Mycoplasma cell well represents an ultimate lower limit for the size of a living organism.

A fundamental problem to understand, West, Brown and Enquist say, is how exchange surfaces and transport distances change, or scale, with body size. In particular, a longstanding question has been why metabolic rate scales as the ¾-power of body mass[76]. Biological scaling can be described by the allometric equation $Y=Y_oM^b$, whereas Y_o varies with the trait and type of organism, b characteristically takes on a limited number of values, all of which are simple multiples of ¼. The question has been why the exponents of the diameters of tree trunks and aorta scale as M 3/8 rates of cellular metabolism and heartbeat as M-1/4, and whole organism metabolic rate as M ¾, are multiples of ¼ rather than 1/3 as expected on the basis of conventional Euclidean geometric scale. The answer is quite simple: a larger surface area gives more capability for electromagnetic absorption, a principle similar to that of antennas.

Proposed models based in fractal-like architecture of the hierarchical branching vascular networks that distributes resources as molecules that eventually will go into endergonic reactions, within organisms accurately predict scaling exponents that have been measured for many structural and functional features of mammalian and plant vascular systems, however it is very difficult to fit with the ubiquitous ¾-power scaling of metabolic rate in diverse kinds of organisms with their wide variety of network designs, and especially in unicellular algae and protists, which have no obvious branched anatomy.

However, a more general model based on the geometry rather than hydrodynamics, and maximizing metabolic capacity by means of the rate at which energy and material resources are taken up from the environment. This is equivalent to maximizing the scaling of whole-organism metabolic rate, B, that is limited by the geometry and scaling behavior of the total effective surface area, a; across which nutrients (e.g. glucose) and energy (sunlight in animals and plants) are exchanged with the external or internal environment. The effective surface area is "maximally fractal" and the network structure is volume-filling. It is in this sense that organism have exploited a fourth spatial dimension[77] by evolving hierarchical fractal-like structures to maximize resource acquisition and allocation. By understanding the harnessing of solar energy by animals, the trend of evolution towards an effective surface is clearly valued.

Viewed from the perspective of human photosynthesis, that provide one great unifying principle; it is no accident, therefore, that many biological networks exhibit area-preserving branching, even though different anatomical designs exploit different hydrodynamic principles. Although living things occupy a three-dimensional space, their internal physiology and anatomy operate as if they were four-dimensional.

Quarter-power scaling laws are perhaps as universal and as uniquely biological as the biochemical pathways of metabolism, the structure and function of the genetic code, and the process of natural selection. The vast majority of organisms exhibit scaling exponents very close to 3/4 for metabolic rate and to 1/4 for internal times and distances. These are the maximal and minimal values respectively, for the effective surface area and linear dimensions for a volume-filling fractal-like network. On the one hand, this is testimony of the power of natural selection, which has exploited variations on this fractal theme to produce the incredible variety of biological form and function. On the other hand, it is testimony to the severe geometric and physical constraints on metabolic processes, which have dictated that all of these organisms obey a common set of quarter-power scaling laws, dictated mainly by the absorption of radiation or size of the wavelength absorbed, that is why it seems like general behavior or rather a general law.

By the early 1930s, physics was a mature science abounding in universally applicable laws mainly because the matter of study of Physics exhibit a behavior and characteristics that are constant in more or less degree. In comparison organismic biology was overwhelmingly descriptive and lacked quantitative expressions that could apply to a broad range of animals or plants because the living thing seems to have a wide diversity. In 1932 Max Kleiber changed all that when he published a paper on "Body size and metabolism" in Hilgardia which included a graph plotting the log of the body weight of mammals against the log of their basal metabolic rate (BMR).

Although this initial data set was rather limited, it contained mammals ranging from rats to steers, a range of body weights spanning three orders of magnitude. As BMR measures energy expenditure at rest, in a post-absorptive state (digestion increases metabolism) and in a thermo neutral environment, it conveys fundamental information about animal´s nutrition needs and allows fascinating intra- and interspecific comparisons.

Hundreds of BMRs are now available for both cold and warm-blooded (exothermic and endothermic) species, and they confirm Kleiber´s ¾ law across 21 orders of magnitude, from unicellular organism to whales. Even more, Knut Schmidt-Nielsen has concluded that the ¾ slope is representative for all ectotherms. And the explanation from the point of view of Human Photosynthesis is quite simple: the ¾ slope obeys to the nature of light absorbed, visible and invisible, because melanin has the capacity to absorb all kinds of energy, the totality of the electromagnetic spectrum, even gravitons. From now on we could and must divide nutrition in two parts: Energy (absorbed from the electromagnetic spectrum, visible and invisible) and building blocks taken from the meals ingested. If we expressed the concept in other words then we have: Glucose is only a source of biomass, water is the source of energy.

Fractal geometry has given us an added dimension, allowing us to understand a bit more of the general principles of biology, however, what is really important for all living beings is optimizing the absorption of photonic energy. Every living thing obeys the rules of scaling discovered by Max Kleiber[78], and in turn Kleiber´s law obeys the general characteristics of electromagnetic radiations and their physicochemical interaction with melanin and water.

Therefore the discovery of Human Photosynthesis, or more exactly, animal photosynthesis, is the ultimate explanation for the longstanding mystery of Kleiber´s law. This presents a step forward in the better understanding why the size of living things is of such fundamental importance.

Chapter 27

The hitherto unknown capacity of melanin to dissociate the water molecule explains the relationship between Aging, Frailty and Cognitive decline.
Or
Aging, Frailty, Cognitive decline and the Origin of Life.

Authors:

Arturo Solís Herrera, MD, PhD[1]**, María del Carmen Arias Esparza MD, MSc[1], Ruth I. Solís Arias, MD[1], Paola E. Solís Arias MD[1], Martha P. Solís Arias, MD[1]

**Corresponding author: Arturo Solís Herrera, MD, PhD
email: comagua2000@yahoo.com
[1] Human Photosynthesis Study Center, López Velarde 108, Centro, Aguascalientes 20000, México.

Key words:

Frailty, Cognitive decline, Sarcopenia, Melanin, Water and Human Photosynthesis, Origin of life.

Being the oldest reaction and therefore, the origin of life, the water dissociation or human photosynthesis will change our perception of many phenomena in our daily live. Of course, the discovery of the hitherto unknown capacity of melanin to dissociate the water molecule not modify the movement of the earth translation neither the moon phases, but the impact in many fields of human knowledge will be enormous. Many aging changes and diseases considered with no treatment have a new real hope.

There has been a 30-fold increase in older people in the United States since 1870, from 1 million up to 35 million in 2000. By 2030, the proportion of the population over age 65 will reach 20 %. Trends in birthrates, death rates, and the flow of cohorts all contribute to population aging ([79]). Individual aging involves changes in biological functioning that are tangible and undeniable. Medical interventions for chronic illness of aged population must be improved. This chapter is organized around controversies, along the facts and basic concepts that stand behind aging.

Aging is a gradual process. The process of aging begins much earlier in life. The biological indicators, or biomarkers, that might identify features of the basic process of aging ([80]) are diastolic and systolic blood pressure, auditory or visual acuity, ability of kidney to excrete urine and the behavior of the immune system; hormones levels, all these tend to decline with chronological age ([81]).

With aging, height generally diminishes while weight increases, skin wrinkles and hair becomes thinner. There is also a loss in vital capacity, or the maximum breathing capacity of the lungs. Both respiratory and kidney functions decrease.

A key finding from studies of biological aging is that chronological age alone is not a good predictor of functional capacity or biological age. People of the same chronological age may differ dramatically in their functional age, which can be measured by biomarkers (⁸²).

Functional capacity in aging is not reduced gradually by one single overall mechanism; however, scientists have come to believe that the process of aging might be controlled at the most basic level of organic life.

If we examine metabolic pathways, and compare biochemical reactions using our critical thinking, we could find the crucial elements. We need to find data that qualifies as basic level of organic life.

Scientific sources agree that the more basic an organic reaction, the more ancient it should be, and the oldest organic reactions are those connected with the harnessing of the sun's energy. Therefore, is worth taking a closer look at Melanin, which up to now was unknown as an energy source.

Melanin is the equivalent of chlorophyll in humans. Both compounds are able to split or dissociate the water molecule.

In a brief comparison between both photo-systems we have:

Compound	Main action	Energy source	Stability	Reaction happens	Abundance in the Universe	Origin of Life
Chlorophyll	Only Dissociation of water molecule	400 and 700 nm of wavelength	20 seconds after it is extracted out of the chloroplast	Strictly inside living things	Light / Water / Chlorophyll	No

Melanin	Dissociation and reformation of water molecule	All kinds of energy	Million years	Inside and outside living things	Light / Melanin / Water	Yes

In a simplified form, the unexpected basic reaction that happens in the human photo-synthesis system (Light/ Melanin/ Water) detected at first in human retina in 1990, and finally deciphered in 2002 ([83]) occurs in the following way:

$$2H_2O \leftrightarrow 2H_2 + O_2 + 4e^-$$

The human photo-system drives the reaction in both senses; however it is not symmetrical in time. Chlorophyll absorbs only red and blue light (400 and 700 nm wavelength), however the capacity of absorption of melanin is the full electromagnetic spectrum (from gamma rays to radio rays), and we might even say that it absorbs any form or kind of energy. Mainly due to its greatest efficiency, the human photosynthesis system can generate enough continuous energy so our body can keep multiple cell work performances 24 hours a day.

Amazingly human body cells can produce energy at their own by means of the relatively simple mechanism of split and reform the water molecule.

Our body begins to lose this important capacity at 26 years old. We lose our capacity to perform the photosynthesis function at its peak at an approximate rate of 10% with each decade of life after the mid-twenties, and when we reach our fifties it goes into free fall. Other factors which negatively affect the process of photosynthesis in humans, besides aging, are: cold weather as in plants happen, iron supplements, drinking alcohol, female hormones, pesticides, herbicides, high fructose syrup, antidepressants, anesthetic agents, a good number of medications, especially those with elevated volume of apparent distribution, contaminated water, chronic poor-light level conditions, life-style and diet.

Our study was started in 1990 looking for new treatments in regards the three main causes of vision loss, that in warm countries are: Glaucoma, diabetic retinopathy, and age-related macular degeneration; and in accordance with the prevalent theories about that blood vessels fills any metabolic requirement of tissues, we expected to identify that very early changes of the mentioned illnesses were originate in endothelial vascular cells, however, they don't.

Our lab equipment was modified in order to get enough magnification and then we could observe any early date about size, direction, or some one of the several anatomic characteristics of the optic nerve capillaries, so we could attempt early medical treatment.

The obtained magnifications were enough to our blood vessels aims, however an unexpected finding changes the variables in study, the ever presence of melanin.

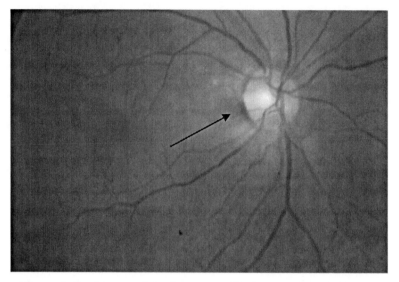

Figure 1. Ocular normal fundus *in vivo*. Black arrow shows melanin.

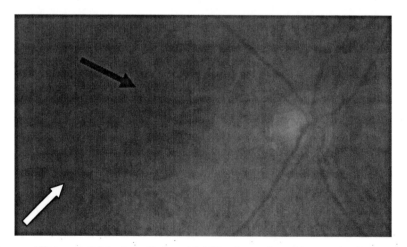

Figure 2. Melanin and choroidal blood vessels and intertwined.
Black arrow: Melanin, white arrow choroidal vessels.

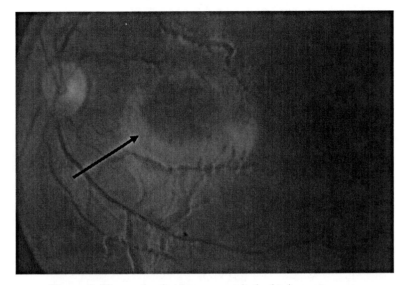

Figure 3. The ocular fundus zone with the highest pigment concentration is in the macula (black arrow).

It is so striking that the Ophthalmology´s text books describe the melanin´s function inside the eye as only a tapered layer whose main function is only the absorption of the light-excess in order to get just a clearer image, with less aberrations or irregular reflections, and didn´t say anymore.

However, as result of the progress of our investigation, the doubts soon begin. If melanin function only is as antireflective coat, why is so constant the melanin presence? In practically all patients studied at that time, around 6000, the "sunscreen" appeared in the optic nerve of every one. In spite that is widely accepted that the amounts of melanin inside the eye have not changes in time, why the melanin located in the rim of the optic nerve changes along the individual life?

Inexorably, melanin changes in most of eye diseases. Why this highly stable molecule with no apparent biological activity beside light absorption reacts with practically every retinal disease?

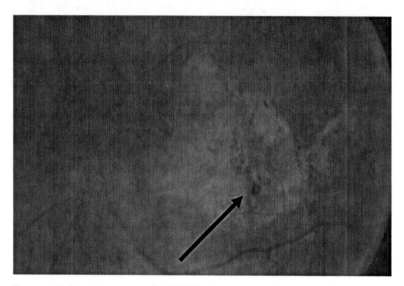

Figure 7. In this case of ARMD there are sub-retinal fibrosis and pigmentation changes. Black arrow: melanin.

Gradually we could understand that melanin is so far of the accepted role. After 12 years of continuous work based in the observation of the ocular fundus *in vivo* in healthy and diseased patients, our doubts finally were cleared in February 2002; Melanin has the unexpected capability to split or dissociate the water molecule. This astonishing finding unravel that the very first reaction in plants is identical in animals. The life begin in both kingdoms with the same process, however, in plants occurs in a very limited form, chlorophyll only can split the water molecule and humans not only can dissociate it, instead our photosystem has the capability to reform the molecule and give us water again depending of factors relatively common in this kind of processes like pressure, temperature; reactants concentrations and so on, as any other chemical reaction that could be reversible.

Briefly, in humans, and in the animal kingdom in general we have:

$$2H_2O \leftrightarrow 2H_2 + O_2 + 4e^-$$

In plants the reaction is very different:

$$2H_2O \rightarrow 2H_2 + O_2$$

At light of our discovery, we were getting gradually that water dissociation or human photosynthesis is a basic reaction not only for the origin of life but for the operation of our body. The fact that the capacity to separate the water begins to reduce at 26 years approximately 10 % each decade and after fifties goes into free fall; is amazingly congruous with the characteristics of aging.

Our body requires the energy that the human photosynthesis releases not only as starting point but every minute is necessary. We need energy constantly not only to stay alive, but to retain the shape, beside many other characteristics that are in a living organism. Every one of the thousands of biochemical reactions that occur in our bodies at every time and with the most various purposes have a common initial requirement: energy.

As the source of initial energy starts to fall the cell´s characteristics developed on the basis of billions of years of evolution are deteriorating inexorably. Then the multiple functions of our bodies are impaired and sooner or later their impact in the quality of life will be significant.

One conclusion of our study is that the three main causes of blindness as the aging process itself have their origin in the loss of efficiency of melanin as transducer, as part of the human photo-system composed by Light / Melanin / Water.

However, that Kleiver published since 1924 his thesis: "The Energy Concept in the Science of the Nutrition", our study shows that until the date there is a persistence of the generalized mistake the belief that food has capability to provide as much energy as nutrients. In accordance with our findings, meals are source of building-blocks, a very important role indeed, but energy is different.

Researchers invest great material and human resources in creating Nutrition-based treatments for the aging, but I disagree. The problem is that nutrient-dense foods lack a common definition. A 1977 review of the literature showed that there were only limited efforts to define the concept of a nutritious food. General statements that such a food should provide "significant amounts of essential nutrients" were not backed by any firm standards or criteria ([84]). Three decades later, in 2004, there was still no agreement as to the definition of a nutrient-dense food or a healthful beverage ([85]).

The concept that "all foods can fit" so there is no reason to emphasize the nutritional quality of individual foods, and the dogma that there are no good or bad foods, only bad diets

([86]), reflects the real or main function of foods: only as a source of carbon chains of different lengths, and orientation, that in combination with other elements, such as Nitrogen, Oxygen, Hydrogen, etc., sustain the body-building mechanisms of our organism.

The Nutrition-based treatments created are spiraling down into a black hole of confusion, paralyzed by what seems to be an increasingly complex and ultimately unmanageable task, that of improving the quality of life in the elderly.

The good news is that most of the biomarkers of aging, frailty, sarcopenia and cognitive decline, could be improved when the understanding that food is only a source of building blocks, and not a source of energy, because energy comes from water, or more precisely, from the photo-system composed by Light/ Melanin/ Water.

Energy is very different to a building block, in the Dave Watson definition: *Energy is a property or characteristic (or trait or aspect?) of matter that makes things happen, or, in the case of stored or potential energy, has the "potential" to make things happen.*

Energy is required to organize those elements named building blocks in a coherent form, fully congruous with the living cell. Human photosynthesis shows us that energy comes from water, not from food.

In light of the knowledge that our body needs energy even to keep its shape, besides other complex uses of energy by the cells and the body, the areas of frailty, sarcopenia, decline in body composition and functionality, and their relationship to cognitive decline and dementia, will have new, different and exciting approaches([87]).

Being the oldest reaction, energy generation through melanin´s water splitting can be considered as the very basic reaction that finally supports the other ones in the long chain of life. Therefore, when this important process is enhanced, the organism tends to the balance as the other biochemical reactions that undoubtedly depend of it. The therapeutic responses are usually dramatic.

In any system, when failure is generalized or disseminated as it occurs in aging, with sarcopenia, frailty, cognitive deficit, Alzheimer's or Parkinson's Disease present, we ought to first think about the role that energy plays in the living organism.

Conflict of interest statement: None.

Chapter 28

The discovery of the intrinsic property of melanin to split and reform the water molecule.

For the observation of the minute blood vessels of the optic nerve and macular zone in the live patient, different wavelengths were used.

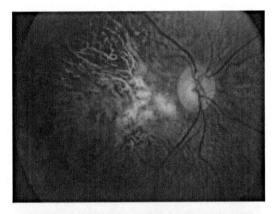

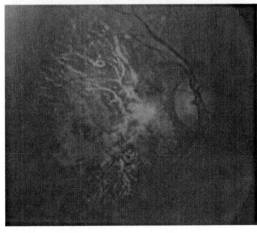

With different wavelengths, details of the observed structures have better contrast.

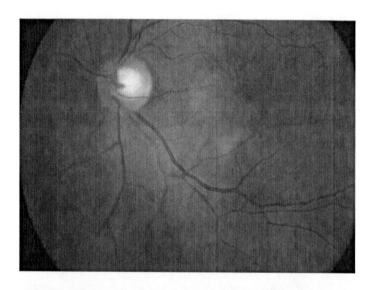

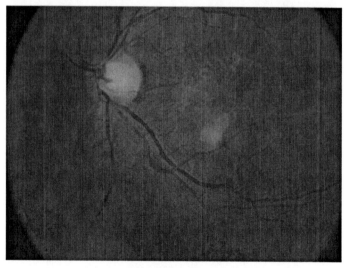

Notice the differences in the macular zone in this case of macular pocket.

Using polychromatic (white) light and blue (monochromatic) light in this case of Ocular Ischemic Syndrome; the damage of the optic nerve and the macular area is well demonstrated.

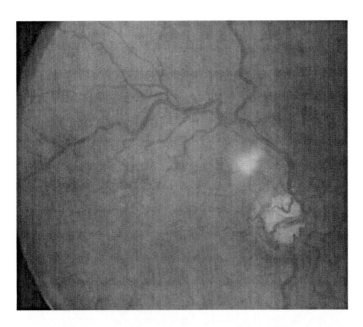

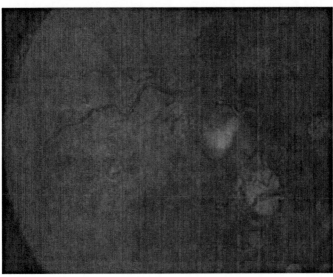

My first success cases were in ophthalmic pathology, these photographs shows the very encouraging results of the enhancement of human photosynthesis in Diabetic retinopathy.

In less of three weeks, the hemorrhage that cover the macular area with the consequent loss of vision, is improved significantly.

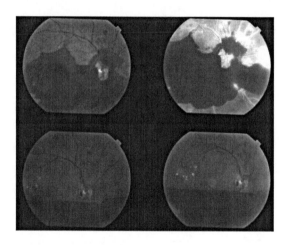

Monochromatic visible light and the study of diabetic retinopathy, shows important retinal details.

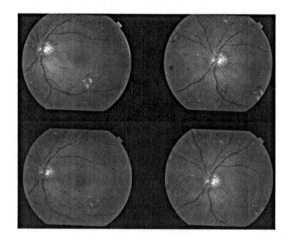

When for illumination of the ocular fundus, invisible light is used, then retina tissue seems as disappear and choroidal tissue is observable easily.

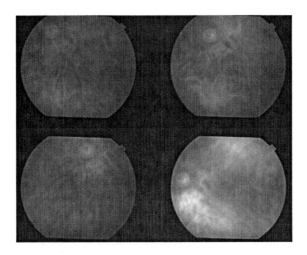

A case of hemorrhage in para-macular area, improved significantly within four weeks of treatment with human photosynthesis enhancement.

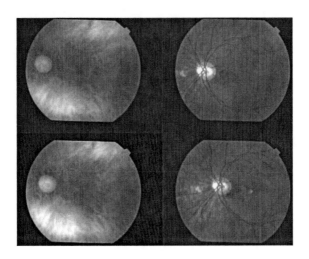

Cerebrovascular events improved significantly with the enhancement of human photosynthesis, tissues affected by bloodshed do not have sequels from hemorrhage. In this case of CVE (subarachnoid hemorrhage) improvement of the optic nerve is impressive.

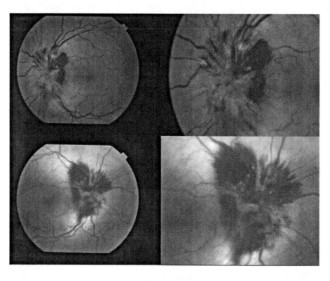

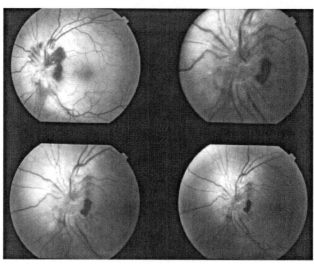

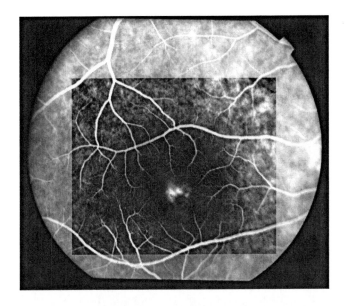

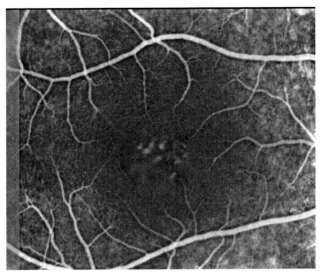

Macular edema responds well to the enhancement of human photosynthesis, in just 10 days of treatment, edema diminished significantly.

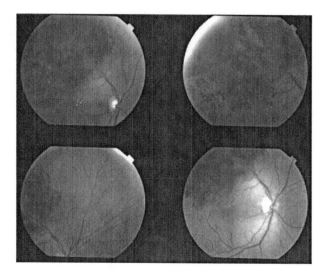

Retinal vascular event, the photograph shows the bloodshed at first examination.

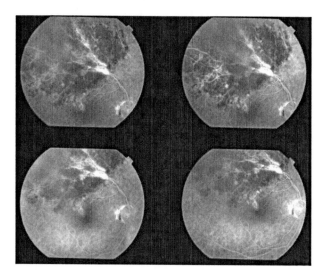

Fluorescein angiography shows the occlusion in an av crossing.

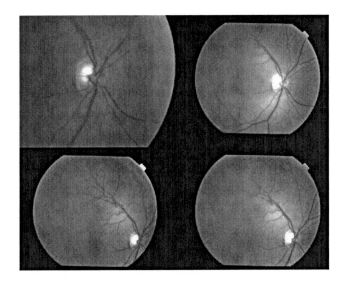

Six weeks later the improvement is amazing.

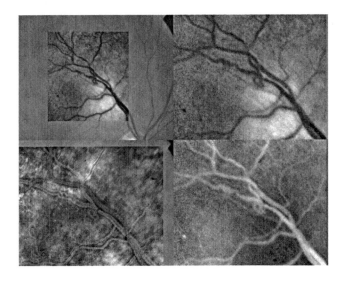

Magnification of the trigger zone, where the vascular event started. There are few vascular alterations, the vessels are recanalized.

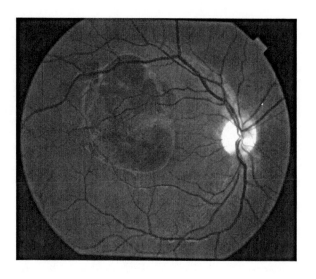

Choroidal hemorrhage after Valsalva´s manoeuvre, the photograph, at first examination.

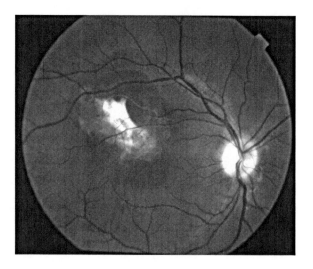

After eight weeks of treatment, with human photosynthesis enhancement, the tissue shows enough recovery that allowed 20/30 vision.

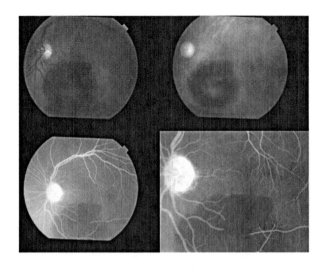

Hemorrhage that covered macular area. With human photosynthesis enhancement, 10 days later, the fovea was recovered.

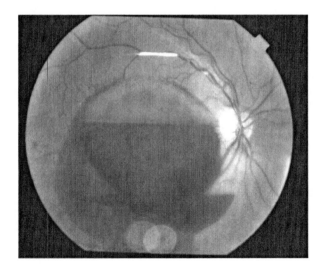

Photograph shows the blood level as the absorption was gradually improving vision.

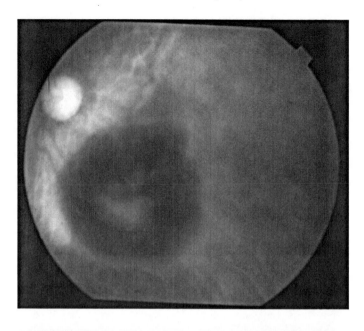

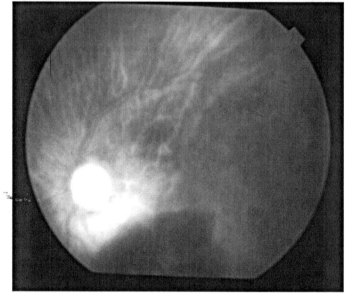

With long wavelength observation, the choroidal blood vessels show changes secondary to the bloodshed.

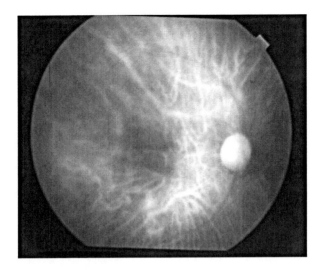

The choroidal blood vessels cannot be examined with white light, therefore long wavelengths must be used, and otherwise choroidal changes cannot be observed adequately.

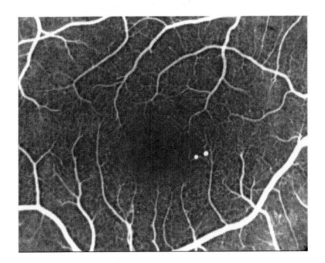

The micro-aneurysm formation, which are characteristic of diabetic maculopathy.

Female patient with macular hemorrhage after prolonged coughing fit. Photographs were taken at first examination.

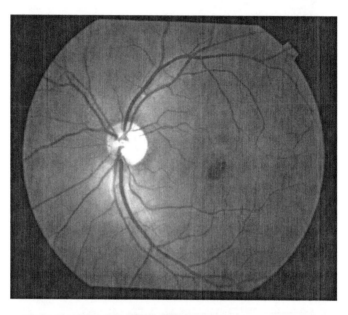

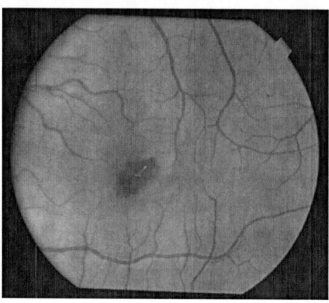

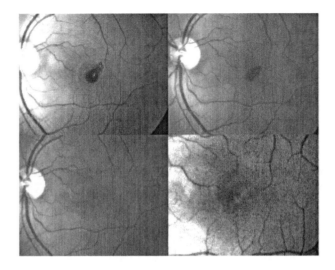

In less of one month, the blood clot disappears, leaving a remarkably healthy macular tissue. Final vision: 20/25.

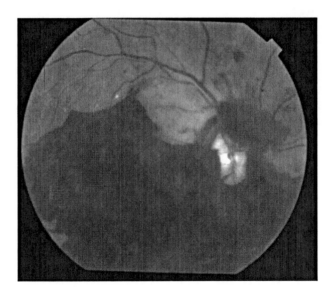

Diabetic patient with an hemorrhage which impairs the macular zone with consequent vision loss.

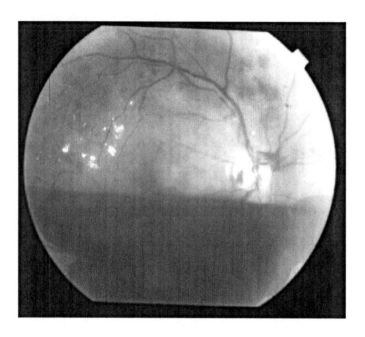

After one week of human photosynthesis enhancement, the macular area now is free with improvement of central vision.

Alzheimer's Disease:

In spite of great technological advancement, the AD patients are dying in the same way that in 1906.

The article from Jean Marion is a vivid description of the patient suffering and relatives:

http://www.slate.com/blogs/quora/2013/02/27/what_is_it_like_w hen_one_of_your_parents_gets_alzheimer_s.html#-1

The photographs of the article are very illustrative:

The author's mother, before and after the ravages of dementia.
Courtesy of the author

The discovery of the human photosynthesis constitutes a real hope for AD patients. Since 2006 until today, around 600 AD patients had been treated at our Human Photosynthesis Study Center. We show only two examples of our extraordinary therapeutic results:

413

Female AD patient, photographed at first examination; 2010.

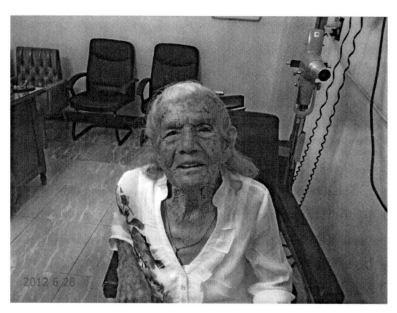

Two years later, the improvement is obvious.

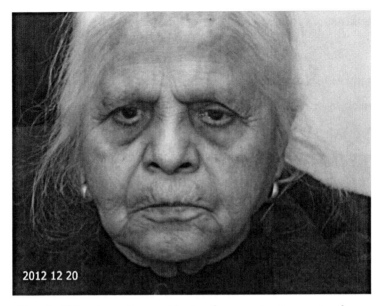

Female AD patient, photographed at first examination, notice date:
2012/12/20.

Five weeks later the improvement is very significant. The date: 2013/1/28

Parkinson´s Disease and Human Photosynthesis:

The main function of melanin in substantia nigra is energy production. It is like a battery that serves as relay station. When the amount of melanin is diminished then the amount of available free chemical energy in mesencephalon is not enough to support the normal functions; for instance: movement.

Some theories say that iron increases when melanin is diminished, however, iron is a strong poison for melanin.

Taking into account that melanin release the energy symmetrically in all directions; then the placement of the substantia nigra is highly strategic. Thereby Melanin is the source of energy for nearby neuronal structures, as cerebral peduncles, red nucleus; etc.

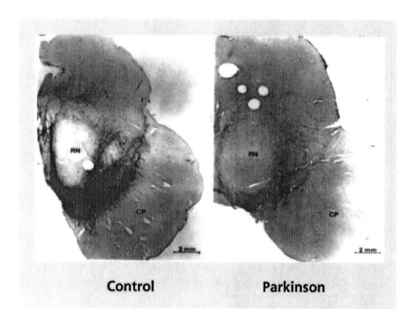

Control   **Parkinson**

The amount of melanin (and therefore energy) is critical for the adequate functioning of mesencephalon. Less melanin means less free chemical available energy. Image source: http://openi.nlm.nih.gov

Probably the loss of melanin is the key point in Parkinson´s Disease.

A few recent observations taken from several sources

- White people have a bigger cardiac frequency than black people (Berenson 1984). I explain this due to their lesser melanin content, thus white people needs more oxygen from the environment to balance the initial reaction which unfolds water.

- Black people has less Hb per liter than white people (1 to 2 gr/dL). This is congruent with energy being produced significantly on the tissue itself, so they require less hemoglobin to carry on to the tissues. It also says lungs are smaller the darker the skin gets.

- With a bigger melanin amount there are lesser GSH and GSH reductase levels (Halprin, Ohkawara, 1966; Benedetto, 1981, 1982) Both systems are antioxidants, but melanin is out of discussion more powerful.

- The whiter the skin gets, there is more risk of systemic arterial hypertension (hypoxia caused vessels constriction. Renine levels are lower on black people).

- The whiter the skin gets, there is more risk of persistent bad eyesight, hypoacusia and several cutaneous carcynoma (Kiyoshi 1996).

- Renal insufficiency risk by any cause is 6 times bigger on black people compared to white people.

And about melanin we have:

- It has the notable ability to absorb the near infrared, visible light and ultraviolet. The reason is unknown.
- It's the only known biopolymer which both in vivo as in vitro contains free radicals center. It has been proved that free radicals concentration can be reversible up to two times its magnitude.
- Melanin has amorphous semiconductor properties.
- It has electron taker properties.

ABOUT THE AUTHOR

Dr. Arturo Solis Herrera was born on Mexico City, on august 19, 1953, coursed elementary education on Luis Hidalgo Monroy school, adjacent to Escuela Nacional de Maestros, he continued his studies at the Prevocational No. 4 from Instituto Politecnico Nacional IPN (National Poly technical Institute) and on Vocational No. 9 Juan de Dios Batiz, also from IPN. The Medicine degree was obtained at the Superior Medicine School of the IPN, ophthalmologist by Universidad Nacional de Mexico (UNAM) and the Conde de Valenciana Hospital. Dr. Solis Herrera studied a specialization on Neuro ophthalmology on Instituto Nacional de Neurologia y Neurocirugia. Got a master's degree on Sciences on Universidad Autonoma de Aguascalientes and, finally, got a PhD in Pharmacology at Universidad de Guadalajara, México.

Mexico hasn't been- until today- a nation characterized by big scientific discoveries, which can change humanity's destiny for the better. However, melanin's photoelectric properties study and electric energy generation with it – by Dr. Arturo Solis Herrera- is amazing and is not wild to think this research could solve some of humankind's energetic challenges on hydrocarbure's substitution and a solution for climatic change.

The finding of a substance worthy of being called "the human chlorophyll", is something really unusual. A substance able to capture light and use its energy to ignite ionic reactions is something only accepted on plants, algae and vegetables' chlorophyll. The discovery of such a mechanism on human beings and on mammals opens a new and promising vein for scientific research.

Alternative and new electric energy generation with melanin charged photo electrochemical cells, separating hydrogen and oxygen from water by electrolysis has been patented by Dr. Solis Herrera. A notable aspect of this cells is they are self-renewable – this means, they don't need to recharge- because once the cell is closed, relentless energy production begins, day and night, for years without molecule decay and without waste of energy and without generating greenhouse's gases.

However, as Dr. Solis Herrera asserts on this book, to massively use this energy generation properties will take an enormous effort that would imply both human and economic investments on a great scale. That's why he and his team are developing applications able to convince the scientific community and the public of the advantages to go deep into this line of research.

Some readers may doubt this is a discovery able to change science and humankind paths, but the most important thing would be the scientific community performing a deep evaluation of this finding and to determine its potential. Besides, we must never lose our ability to amaze ourselves.

David Shields
CEO
Energía a Debate
Magazine

REFERENCES

[1] Scharf, Caleb. The benevolence of the black holes. Scientific American, August 2012, pp. 22-27

[2] Bjerrum, N. 1952. Structure and properties of ice. Science 115:385–390.

[3] Franks, F. 1985. Biophysics and Biochemistry at Low Temperatures. Cambridge University Press. Franks, F. 2000.

[4] Daniel, R. M., J. L. Finney, and A. M. Stoneham, Eds. 2004. The molecular basis of life: is life possible without water? Philos. Trans. R. Soc. London B 359:1141–1328.

[5] Halle, B. 2004. Protein hydration dynamics in solution: a critical survey. Philos. Trans. R. Soc. London B 359: 1207–1224.

[6] Denisov, V. P., and B. Halle. 1996. Protein hydration dynamics in aqueous solution. Faraday Discuss. 103: 227–244.

[7] Sciortino, F., A. Geiger, and H. E. Stanley. 1991. Effect of defects on molecular mobility in liquid water. Nature 354:218–221.

[8] Buckingham, A. D. 1986. The structure and properties of a water molecule. In Water and Aqueous Solutions, eds. G. W. Neilson and J. E. Enderby. Bristol: Adam Hilger.

[9] Tanford, C., The Hydrophobic Effect, 2nd ed. New York, NY: Wiley, 1980. 7. Kauzmann, W. Some factors in the interpretation of protein denaturation. Adv. Protein Chem. 14, 1959:1–63

[10] Lodish H, Berk A, Zipursky SL, et al. Molecular Cell Biology. 4th edition. New York: W. H. Freeman; 2000.

[11] 1 atomic mass unit = $1.66053886 \times 10^{-27}$ kilograms

[12] Lange, Neale R. Schuster, Daniel P. The measurement of lung water. Crit Care 1999, 3: R19-R24

[13] Belding, David L., The respiratory Movements of Fish as an indicator of a toxic environment. Transactions of the American Fisheries Society. Vol. 59, 1929 Iss.10.1577/1548-8659(1929)59[238:TRMOFA]2.0.CO;238-245

[14] Karp, Gerald. In the chapter Introduction to the cell biology, of the book: Cell and Molecular Biology. John Wiley and Sons, Ed. New York, Second Edition. 1996. Page 5.

[15] Sherwood, Lauralee. In the Chapter 3 The plasma Membrane and Membrane Potential of the book: Human Physiology, from Cells to Systems. Thomson, Brooks/Cole, Ed. Australia, Canada, México. Fifth Edition, 2004. Page 78

[16] Takeda E, Taketani Y, Sawada N, Sato T, Yamamoto H. The regulation and function of phosphate in the human body. Biofactors. 2004;21(1-4):345-55.

[17] Scalettar, Bethe A., Abney, James R. Hackenbrock, Charles R. Dynamics, structure and function are coupled in the mitochondrial matrix. Proc. Natl. Acad. Sci. USA. Vol. 88, pp 8057-8061. September 1991, Cell Biology.

[18] Sheltzer, Jason M., Blank, Heidi M., Pfau Sarah J. et al. Aneuploidy Drives Genomic Instability in Yeast. Science 19 August 2011, Vol. 333. Pages 1026-1030

[19] Gunn TR, Easdown J, Outerbridge EW, Aranda JV. Risk Factors in retrolental fibroplasia. Pediatrics. 1980 Jun;65&6):1096-100.

[20] Saunders RA, DOnahue ML, Christmann AV, Tung B, Hardy RJ, Phelps. Racial variation in retinopathy of Prematurity Cooperative Group. Arch Ophthalmol. 1997 May;115(5):604-8.

[21] Sudipta Paul. Hypothesis: fetal growth. BJOG Vol. 106, issue 7; pp749-750, July 1999.

[22] www.webwlwmwnts.com/hydrogen/biology.html

[23] J.A. Kerr in CRC Handbook of Chemistry and Physics 1999-2000: A Ready-Reference Book of CHemical and Physical Data. D.R. Lide, (ed), CRC Press, Boca Raton, Florida, USA, 81st edition, 2000.

[24] J. van Kranendonk and H.P. Gush, *Physics Letters A*, 1962, 1, 22.

[25] Atkins, P. W. Physical Chemistry, 7th ed, Oxford University Press, 2006.

[26] Wilson, J. M. 1887. Essays and Addresses: An Attempt to Treat Some Religious Questions in a Scientific Spirit. London: Macmillan, pp. 1–30.

[27] McKie, D., and N. H. de V. Heathcote. 1935. The Discovery of Specific and Latent Heats. London: E. Arnold. [21] Hope, T. C. 1805. Experiment on the contraction of water by heat. Trans. R. Soc. Edinb. 5:379.

[28] Wilson, J. M. 1887. Essays and Addresses: An Attempt to Treat Some Religious Questions in a Scientific Spirit. London: Macmillan, pp. 1–30.

[29] Williams, R. J. P., and J. R. R. Frausto da Silva. 2006a. Evolution revisited by inorganic chemists. In Fitness of the Cosmos for Life: Biochemistry and Fine-Tuning, ed. J. D. Barrow, S. Conway Morris, S. J. Freeland, and C. L. Harper Jr. Cambridge: Cambridge University Press.

[30] Henderson, L. J. 1917. The Order of Nature. London: Harvard University Press. [70] Mason, F., Ed. 1934. The Great Design: Order and Progress in Nature. London: Duckworth. [71] Ibid., pp. 187–206.

[31] Pimentel, G. C., and A. L. McClennan. 1960. The Hydrogen Bond. San Francisco, CA: Freeman. [84] Russell, C. A., 1975. A drop in the ocean. In Wonders of Creation, ed. Pearman, R., M. Fergus, and P. Alexander. Berkhamsted: Lion, pp. 77–81.

[32] Denton, M. J. 1998. Nature's Destiny. New York, NY: The Free Press, p. 22.

[33] Denton, M. J. 1998. Nature's Destiny: How the Laws of Biology Reveal Purpose in the Universe. New York, NY: The Free Press, pp. 19–46.

[34] Bjerrum, N. 1952. Structure and properties of ice. Science 115:385–390.

[35] Franks, F. 1985. Biophysics and Biochemistry at Low Temperatures. Cambridge University Press. Franks, F. 2000.

[36] Daniel, R. M., J. L. Finney, and A. M. Stoneham, Eds. 2004. The molecular basis of life: is life possible without water? Philos. Trans. R. Soc. London B 359:1141–1328.

[37] Halle, B. 2004. Protein hydration dynamics in solution: a critical survey. Philos. Trans. R. Soc. London B 359: 1207–1224.

[38] Denisov, V. P., and B. Halle. 1996. Protein hydration dynamics in aqueous solution. Faraday Discuss. 103: 227–244.

[39] Sciortino, F., A. Geiger, and H. E. Stanley. 1991. Effect of defects on molecular mobility in liquid water. Nature 354:218–221.

[40] Buckingham, A. D. 1986. The structure and properties of a water molecule. In Water and Aqueous Solutions, eds. G. W. Neilson and J. E. Enderby. Bristol: Adam Hilger.

[41] Tanford, C., The Hydrophobic Effect, 2nd ed. New York, NY: Wiley, 1980. 7. Kauzmann, W. Some factors in the interpretation of protein denaturation. Adv. Protein Chem. 14, 1959:1–63

[42] Maréchal, Yves. The Hydrogen Bond and the Water Molecule. Elsevier, Amsterdam, Boston, Heidelberg, London, New York. First Edition 2007, Page 23.

[43] Maddox, John. Crystals from first principles. *Nature* 335, 201 (15 September 1988) | doi:10.1038/335201a0

[44] Erich Sackmann, *Biological Membranes Architecture and Function.*, Handbook of Biological Physics, (ed. R.Lipowsky and E.Sackmann, vol.1, Elsevier, 1995)

[45] Hillman, Robert S.; Ault, Kenneth A.; Rinder, Henry M. (2005). *Hematology in Clinical Practice: A Guide to Diagnosis and Management* (4 ed.). McGraw-Hill Professional. p. 1. ISBN 0071440356

[46] Pierigè F, Serafini S, Rossi L, Magnani M (January 2008). "Cell-based drug delivery". *Advanced Drug Delivery Reviews* 60 (2): 286–95. doi:10.1016/j.addr.2007.08.029. PMID 17997501

[47] Kleinbongard P, Schutz R, Rassaf T, et al (2006). "Red blood cells express a functional endothelial nitric oxide synthase". *Blood* 107 (7): 2943–51. doi:10.1182/blood-2005-10-3992. PMID 16368881

[48] Kabanova S, Kleinbongard P, Volkmer J, Andrée B, Kelm M, Jax TW (2009). "Gene expression analysis of human red blood cells". *International Journal of Medical Sciences* 6 (4): 156–9. PMC 2677714. PMID 19421340

[49] Iron Metabolism, University of Virginia Pathology. Accessed 22 September 2007

[50] Muñoz V, WL BUTLER 1975 Photoreceptor pigment for blue light in Neurospora crassa.Plant Physiol 55: 421-426

[51] ROBERT D. BRAIN, JOHN A. FREEBERG,3 CHARLES V. WEISS,4 AND WINSLOW R. BRIGGS. Blue Light-induced Absorbance Changes in Membrane Fractions from Corn and Neurospora. Plant Physiol. (1977) 59, 948-952

[52] Aloia John F. African Americans, 25-hydroxyvitamin D, and osteoporosis: a paradox. Am J Clin Nutr August 2008 vol. 88 no. 2 545S-550S

[53] Schwarz, Michael; Jerzey, Bill. Fractals, Hunting the Hidden Dimension. Nova. ISBN 978-1-59375-852-3

[54] Schmidt-Nielsen, Knut. Scaling, why is Animal Size so Important. Cambridge University Press. 1984. ISBN 0 521 26657 2

[55] West, Geoffrey B., Brown, James H. The origin of allometric scaling laws in biology from genomes to ecosystems: towards a quantitative unifying theory of biological structure and organization. J Exp Biol 208, 2005; 1575-1592. Doi: 10.242/Feb. 01589

[56] West, G. B., Woodruff, W. H. and Brown, J. H. (2002b). Allometric scaling of metabolic rate from molecules and mitochondria to cells and mammals. *Proc. Natl. Acad. Sci. USA* 99, 2473-2478.

[57] Reich PB, Tjoelker MG, Machado JL, Oleksyn J (26 January 2006). "Universal scaling of respiratory metabolism, size, and nitrogen in plants". *Nature* 439 (7075): 457–61. doi:10.1038/nature04282. PMID 16437113

[58] Wang, Zimian., O´Connor, Timothy P., et al. The Reconstruccion of Kleiber Law at the Organ-Tissue Level. J Nutr. 2001 Nov;131(11):2967-70.

[59] Hall, C. W. (1999) Laws and Models: Science, Engineering, and Technology. CRC Press, New York.

[60] 1. Smil, V. of scaling discovered by Max Kleiber. Nature 403: 597. (2000) Laying down the law: every living thing obeys the rules

[61] Hulbert, A. J. & Else, P. L. (2000) Mechanisms underlying the cost of living in animals. Ann. Rev. Physiol. 62: 207–235.

[62] West, G. B., Brown, J. H. & Enquist, B. J. (1999). The fourth dimension of life: fractal geometry and allometric scaling of organisms. Science 284:1677–1679.

[63] Abraham Szoke, David van der Spoel and Janos Hajdu, Energy Utilization, Catalysis and Evolution-Emergent Properties of Life, *Current Chemcial Biology*, 2007, *1*, 53-57.

[64] Jensen, F. Introduction to computational chemistry, Chichester, Wiley & Sons (1999).

[65] Proctor, PH., McGinnes J.E., Arch Dermatol 122:507-508, 1986.

[66] Alberts B, Johnson A, Lewis J, Raff M, Roberts K, Walter P. Molecular Biology

of the Cell (4th edition). New York, Garland (2002).

[67] Berg JM, Tymoczko JL, Stryer L. Biochemistry (5th edition) New York, Freeman (2002).

[68] Szoke A, Scott WG, Hajdu J. Catalysis, evolution and life. FEBS Lett 2003; 553: 18-20.

[69] Laughlin RB, Pines D. The theory of everything. Proc Natl Acad Sci USA 2000; 97: 28-31.

[70] Nicholls DG, Ferguson SJ. Bioenergetics 3. San Diego, Academic Press (2002).

[71] de Meis L. Uncoupled ATPase activity and heat production by the sarcoplasmic reticulum Ca2+-ATPase-Regulation by ADP. J Biol Chem 2001; 276: 25078-87.

[72] Halonen P, Baykov AA, Goldman A, Lahti R, Cooperman BS. Single-turnover kinetics of Saccharomyces cerevisiae inorganic pyrophosphatase. Biochemistry 2002; 41: 12025-31.

[73] Arturo Solís-Herrera, María del Carmen Arias Esparza, Ruth I. Solís-Arias, Paola E. Solís-Arias, Martha P. Solís-Arias. The unexpected capacity of melanin to dissociate the water molecule fills the gap between the life before and after ATP. *Biomedical Research 2010; 21 (2):224-226*

[74] Solís-Herrera Arturo, Lara María E., Rendón Luis E. Photoelectrochemical Properties of Melanin. Nature Precedings : hdl:10101/npre.2007.1312.1 : Posted 12 Nov 2007.

[75] West, Geoffrey B., Brown, James H., Enquist, Brian J. The Fourth Dimension of Life: Fractal Geometry and Allometric Scaling of Organisms. Science, Vol 284, 4 June 1999 pages 1677-1679.

[76] K. J. Niklas, *Plant Allometry: The Scaling of Form and Process* (Univ. of Chicago Press, Chicago, IL, 1994).

[77] J. J. Blum, Journal of Theoretical Biology, 64, 599 (1977).

[79] Moody, Harry M. Aging. Pine Forge Press 2010, 6th edition. Sage Publications, LTD, London; United Kingdom.

[80] Shock, N. (1962). The physiology of aging. Scientific American, 206, 100-110.

[81] Warner, H. R. 2004. Current status of efforts to measure and modulate the biological rate of aging. / Gerontol. A Biol. Sci. Med. Sci. 59:692-696.

[82] Anstey, K. J., Lord, S. R., & Smith, G. A. (1996). Measuring human functional age: A review of empirical findings. Experimental Aging Research, 22, 245-266.

[83] Arturo Solís-Herrera, María del Carmen Arias Esparza C., Ruth I. Solís-Arias, Paola E. Solís-Arias, Martha P. Solís-Arias. The unexpected capacity of melanin to dissociate the water molecule fills the gap between the life before and after ATP. *Biomedical Research 2010; 21 (2):*

[84] Guthrie H. Concept of a nutritious food. J AmDiet Assoc 1977; 71:14– 19

[85] Lackey CJ, Kolasa KM. Healthy eating: defining the nutrient quality of foods. Nutr Today 2004; 39:26 –9.

[86] Guthrie HA. There's no such thing as "junk food," but there are junk diets. Healthline 1986;5:11–2

[87] Solís Herrera, Arturo; Arias Esparza, María del Carmen; Alvarado Esquivel, J. Jesús. The pharmacologic intensification of the water dissociation process, or Human Photosynthesis, and its effect over the recovery mechanisms in tissues affected by bloodshed of diverse etiology. IJCM 2011 Vol.2 No.3

Suggested readings:

(1)[87] Iida Tomohiro, Hagimura Norikazu, Sato Taku, Kishi Shoji. Optical Coherence Tomographic Features of Idiopathic Submacular Choroidal Neovascularization. Am J Ophthalmol, Dec 2000, Vol. 130, No. 6. pp 763-768.

(2)[87] Castilla Serna Luis, Jurado García Eduardo, Lares Yassef Ismael, Ruiz Maldonado Ramón, Cravioto Muñoz Joaquín. Propuesta de un instrumento guía para la estimación del riesgo ético en los protocolos de investigación en la que participan seres humanos. Gaceta Médica de México. Vol. 122, Núm. 14. Marzo – abril 1986.

[87](3) Cynthia L Arfken, Ana E. Salicrup, Stacy M. Meuer, Lucian V. del Priore, Ronald Klein, Janet B. Mcgill, Cheryl S. Rucker, Neil H. White, Julio V. Santiago. Retinopathy in African Americans and Whites With Insulin-Dependent Diabetes Mellitus. Arch Intern Med. 1994; 154:2597-2602.

(4)[87] Roy monique S. Diabetic Retinopathy in African Americans With Type 1 Diabetes: The New Jersey 725. Arch Ophthalmol Vol. 118, Jan 2000.

(5)[87] Islas Andrade Sergio, Lifshitz Guinzberg Alberto. Diabetes Mellitus, segunda edición. McGraw-Hill Interamericana 1999, México.

(6)[87]Moss Scot E., Klein Ronald, Meuer Moneen B., Klein Barbara E.K. The Association of Iris Color with Eye Disease in Diabetes. Ophthalmology 94:1226-1231, 1987.

(7)[87] Berenson Gs., Webber LS., Srinivasan SR, Cresanta JL, Frank GC, Farris RP. Black-white contrasts as determinants of cardiovascular risk in childhood: precursors of coronary artery and primary hypertensive diseases. Am Heart J 108(3Pt2):672-83, 1984 sept.

(8)[87] Brancati FL., Whittle JC., Whelton PK., Seidler AJ., Klag MG., The excess incidence of diabetic end-stage renal disease among blacks. A population based study of potential explanatory factors. JAMA 268 (21):3079-84, 1992, December 2.

(9)[87] Materson Barry J., Preston Richard A. Angiotensin-Converting Enzime Inhibitors in Hypertension: a Dozen Years of Experience. Arch. Intern Med. Volume 154 (5), March 14, 1994; pp 513-523.

(10)[87] Wilson Peter. Lipids, Lipases, and Obesity: Does Race Matter?. Aterioescl, Thrombos and Vasc Biol. Vol 20 (8) August 2000, pp 1854 – 1856.

(11)[87] Margo Curtis, Mulla Zuber, Malignant Tumors of the Eyelid. Arch Ophthalmol vol 116, feb 1998

(12)[87] Dante J. Pieramici, Neil M. Bressler, Susan B. Bressler, Andrew P. Schachat. Choroidal Neovascularization in Black Patients. Archives of Ophthalmology on CD-ROM 1994 Vol. 112 August

(13)[87]Moore, P.W. Blue Eyes, Big Earplugs. Science News. Oct 31, 1998.

(14)[87] Richard A. Saunders, MD Margaret L. Donahue, MD Linda M. Christmann, MD Al V. Pakalnis, MDBetty Tung, MS Robert J. Hardy, PHD Dale L. Phelps, MD. Racial variation in retinopathy of prematurity. Arch Ophthalmol. 1997;115:604-608

(15)[87] Quevedo Walter C., Holstein Thomas J., General Biology of Mammalian Pigmentation in the book The Pigmentary System, Physiology and Pathophysiology. Edited by Nordlund James J., Boissy Raymond E., Hearing Vincent J., King Richard A. 1998, Oxford University Press. Pp 43-57

(16)[87] Skolnick Andrew A. Novel Therapies to prevent Diabetic Retinopathy.JAMA Vol 278, Nov 12, 1997.

(17)[87] Lüscher Thomas F., Barton Mathias. Biology of the Endothelium. Clin. Cardiol, Vol 20 (Suppl. II) II-3-II-10 (1997).

(18)[87] Loyd Ricardo V., Scheithawer Bern W., Kuroki Takao, Vidal Sergio, Kovacs Kalman, Stefaneau Lucia. Vascular Endotelial Growth Factor (VEGF) Expression in Human Pituitary Adenomas and Carcinomas. Endocrine Pathology, Vol. 10, no 3, 229-235; Fall 1999.

(19)[87] Moss Scot E., Klein Ronald, Meuer Moneen B., Klein Barbara E.K. Tha association of Iris Color with Eye Disease in Diabetes. Ophthalmology 94:1226-1231, 1987.

(20)[87] Kohner Eva, Aldington Stephen, Sttraton Irene, Manley Susan, Holman Rury, Matthews David, Turner Robert. United Kingdom Prospective Diabetes Study. Arch Ophtahlmol /VOL 116, march 1998, 297- 303.

(21)[87] Panamerican Society for Pigment cell research. Vol 6 number 4, december 1998)

(22)[87] Skolnick Andrew A. Novel Therapies to Prevent Diabetic Retinopathy. JAMA Vol 278, Nov 12, 1997.

(23)[87] Chibber R., Molinatti P.A., Konher E.M., Intracellular protein glycation in cultured retinal capillary pericytes and endothelial cells exposed to high glucose concentration. Cell Mol Biol (Noisy-le-Grand) 1999 Feb; 45 (1): 47-57.

(24)[87] McNeill John H., Experimental Models of Diabetes, Mosby, Toronto, 1998, 128-129

(25)[87] Klein Ronald, Klein Barbara, Moss Scot. Epidemiology of Proliferative Diabetic Retinopathy. Diabetes Care, Vol 15, number 12, December 1992.

(26)[87] (Gillow, J.T; Gibson J M; Dodoson, PM; Hypertension and diabetic retinopathy – what's the story. BJO Vol 83(9) september 1999, pp 1083-1087

(27)[87] Herman William, Eastman Richard C., The effects of Treatment on the Direct Costs of Diabetes. Diabetes Care 21 (suppl. 3) : C19 – C 24, 1998.

(28)[87] Harris Maureen I. Diabetes in America: Epidemiology and Scope of the problem. Diabetes Care, Volume 21, Supplement 3, December 1998. C11-C14.

(29)[87] Haffner SM, Fong D, Stern MP, Pugh JA, Hazuda HP, Patterson JK, Van Heuven WAJ., Klein R, Diabetic Retinopathy in Mexican – Americans and non-hispanic whites.*Diabetes*, 37. 878 – 84, 1978

(30)[87] Cynthia L Arfken, Ana E. Salicrup, Stacy M. Meuer, Lucian V. del Priore, Ronald Klein, Janet B. Mcgill, Cheryl S. Rucker, Neil H. White, Julio V. Santiago. Retinopathy in African Americans and Whites With Insulin-Dependent Diabetes Mellitus Arch. Intern Med, Vol 154 (22) November 28, 1994, pp 2597-2602.

(31)[87] Emily Y. Chew, MD, Michael L. Klein, MD, Frederick L. Ferris III, MDNancy A. Remaley, MS, Robert P. Murphy, MD, Kathryn ChantryByron J. Hoogwerf, MD, Dayton Miller, PHD, Association of Elevated Serum Lipid Levels With Retinal Hard Exudate in Diab The Archives of Ophthalmology on CD-ROM, 1996 Vol. 114 September

(32)[87] Moussa B.H. Youdim, Riederer Peter. Understanding Parkinson 's Disease.Scientific American, January 1997

(33)[87] Dante J. Pieramici, Neil M. Bressler, Susan B. Bressler, Andrew P. Schachat. Choroidal Neovascularization in Black Patients. Archives of Ophthalmology on CD-ROM 1994 Vol. 112 August

(34)[87] Brancati FL., Whittle JC., Whelton PK., Seidler AJ., Klag MG., The excess incidence of diabetic end-stage renal disease among blacks. A population based study of potential explanatory factors. JAMA 268 (21):3079-84, 1992, December 2

(35)[87] Margo Curtis, Mulla Zuber, Malignant Tumors of the Eyelid. Arch Ophthalmol vol 116, feb 1998.

[87] Wilson, Peter W.F. Lipids, Lipases, and Obesity: Does RACE Matter? Arterioscler Thromb Vasc Biol, Volume 20(8), August 2000, 1854-1856.

(36)[87] He Jiang, Klag, Michael J., Caballero Benjamin, Appel Lawrence J., Charleston Jeanne, Whelton Paul K Plasma Insulin Levels and Incidence of Hypertension in African-Americans and Whites. Arch Intern Med, Volume 159(5), March 8, 1999, 498 – 503.

(36)[87] King, George L., Suzuma Kiyoshi. Pigment-Epithelium-Derived-Factor. A Key Coordinator of Retinal Neuronal and Vascular Functions. Clinical

Implications of Basic Research. The New England Journal of Medicines. Vol 342(5), Feb 2000, pp 349-351.

(37)[87] Scott E. Moss, Klein Ronald, Meuer Moneen B., Klein Barbara E.K. The Association of Iris Color With Eye Disease in Diabetes. Ophthalmology, October 1987, Vol 94, Num 10, pp1226-1231.

(38)[87] Glaser BM, Campochiaro PA, Davis JL, Sato M. Retinal pigment epithelial cells release an inhibitor of neovascularizacion. Arch Ophthalmol 1985;103;1870-1875.

(39)[87] Miller H, Miller B, Ryan SJ, The role of retinal pigment epithelium in the involution of subretinal neovascularization . *Invest Ophthalmol Vis Sci* 1986;27:1644-1652.

(40)[87] Morse LS, Terrell J, Sidikaro Y. Bovine retinal pigment epithelium promotes proliferation of choroidal endothelium in vitro. *Arch Ophthalmol* 1989;107:1659-1663.

(41)[87] Thomas F. Luscher, Mathias Barton. Biology of the endothelium.Clin Cardiol. Vol 20 (Suppl. II) II-3-II-10 (1997

(42)[87] Calles-Escandon Jorge, Cipolla Marilyn. Diabetes and Endothelial Dysfunction: a Clinical Perspective. Endocrin Rev 22 (1) : 36-52. 2001.

(43)[87] Meade Tollin, Linda C., Biological Basis for Antiangiogenic Therapy. Acta Histochem 102, 117-127 (2000).

(44)[87] H Saraux, C Lemason, H Offret, G. Renard, Anatomía e histología del ojo. Edit. Masson. París, Barcelona, segunda edición 1985.

(45)[87] Medrano E. Esthela. Aging, Replicative Senescence, and the Differentiated Function of the Melanocyte. In the book: The Pigmentary System, Pathology and Physiology. Edited by James L. Nordlund, Oxford University Press. 1998. New York. Pp 151-158.

(46)[87] Lambert Cook Jonathan, Dzubow Leonard M. Aging of the Skin, implications for cutaneous Surgery. Arch Dermatol 1997, Vol 133; october.

(47)[87] Comunicación personal Rodolfo Nicolaus (http://www.tightrope.it/nicolaus).

(48)[87] Regan Susan, Judge Heidi, Gragoudas Evangelous, Egan Kathleen. Iris Color as a Prognostic factor in ocular melanoma. Arch Ophthalmol VOL 117, june 1999.811-814

(49)[87] Lee Patricia, Wang Cindy, Adamis Anthony. Ocular Neovascularization: An Epidemiologic Review. Survey of Ophthalmology, volume 43, number 3, nov-dec 1998. 245-269.

(50)[87] Richard A. Saunders, MD Margaret L. Donahue, MD Linda M. Christmann, MD Al V. Pakalnis, MDBetty Tung, MS Robert J. Hardy, PHD Dale L. Phelps, MD. Racial variation in retinopathy of prematurity. Arch Ophthalmol. 1997;115:604-608

(51)[87] Dawson DW, Volpert O.V., Gillis P, Crawford S.E., Xu H.J., Benedict W., Bouck N.P., Pigment Epithelium-Derived factor: A potent Inhibitor of Angiogenesis. Science 9 july 1999, vol 285, 245-248.

(52)[87] Halaban Ruth. The Regulation of Normal Melanocyte Proliferation. Pigment Cell Res 13:4-14,2000.

(53)[87] Wilkerson Craig L., Nasreen A. Syed, Marian R. Fischer, Nancy L. Robinson, Ingolf H.L. Wallow, Daniel M. Albert. Melanocytes and iris color, Ligth microscopic findings. Arch Opthalmol. Vol 114 (4), april 1996, pp 437-442.

(54)[87] Kosower Evie, Ernst Allan, Taub Bonnie, Berman Nancy, Andrews Jennifer, Seidel James. Tobacco Prevention Education in a Pediatric Residency Program. Arch of Pediatr & Adolesc Med. Vol 149 (4) April 1995, pp 430 – 435.

(55)[87] Bito Laszlo, Matheny Adam, Cruickshanks, Nondahl David, Carino Olivia. Eye color Changes past early Childhood. Arch Ophthalmol Vol 115, may 1997.

(56)[87] Klein Ronald, Barbara E.K. Klein, Moss Scot E., Cruickshanks Karen J. The Wisconsin Epidemiologic Study of Diabetic Retinopathy. XIV. Ten-year incidence and Progression of Diabetic Retinopathy. Arch Ophthalmol /vol 112, sept 1994

(57)[87] Yip R. Significance of an abnormally low or high hemoglobin concentration during pregnancy: special consideration of iron nutrition. Am J Clin Nutr 2000 Jul, 72 (1 Suppl), 272S-279S.

(58)[87] Scholl TO, Reilly T. Anemia, Iron and pregnancy outcome. J Nutr 2000 Feb; 130(2S Suppl): 443S-447 S.

(59)[87] Kobrinsky NL, Yager JY, Cheang MS, Yatscoff RW, Tenenbein M. Does iron deficiency raise the seizure threshold?. J Child Neurol 1995 Mar; 10(2):105-9.

(60) Jose Neptuno Rodriguez-Lopez$, Jose Tudelap, Ramon VaronS, Francisco Garcia-Carmonap and Francisco Garcia-Canovaspll Analysis of a Kinetic Model for Melanin Biosynthesis Pathway THE JOURNAL OF BIOLOGICAL CHEMISTRY 0 1992 by The American Society for Biochemistry and Molecular Biology, Inc Vol. 261, No. 6, Issue of February 25, pp. 3801-3810,1932. *Printed in U. S. A.* (Received for publication, October 11, 1991)

(61)Shosuke Ito Reexamination of the structure of eumelanin. School of Hygiene, Fujita-Gakuen Health University, Toyoake, Aichi 470-11 (JapanBiochimica et Biophysica Acta 883 (1986) 155-161 155 Elsevier BBA 22416

(62) Helene Z. Hill The Function of Melanin or Six Blind People Examine an Elephant. BioEssays Vol 14, No 1 – January 1992, 49.

(63)Y. T. THATHACHARI and M. S. BLOIS PHYSICAL STUDIES ON MELANINS II. X-RAY DIFFRACTION BIoPHYsICAL JOURNAL VOLUME 9 1969

(64)DANIEL THOMAS HARRIS XXXVIII. OBSERVATIONS ON THE VELOCITY OF THE PHOTO-OXIDATION OF PROTEINS AND AMINO-ACIDS. From the Departments of Physiology and Biochemistry, University College, London. (Received February 25th, 1926.)

65. Ito Seigo, Murakami Takurou N, Comte Pascal, Liska Paul, Grätzel Carole, MohMMAD k. Nazeeruddin, Michael Grätzel. Fabrication of thin film dye sensitized solar cells with solar to electric power conversión efficiency over 10 %. Thin Solid Films 516(2008), 4613-4619.

66. Hill Helene Z. The function of Melanin or six Blind People Examin an Elephant. BioEssays, Vol 14, Number 1, January 1992.

67. Songyuang Dai, Jian Weng, Yifeng Sui, Shuanghong Chen, Shangfeng Xiao, Yang Huang, Fantai Kong, Xu Pan, Linhua Hu, Changneng Zhang, Kongjia Wang. The design and outdoor application of dye-sensitized solar cells. Inorganica Chimica Acta 361 (2008) 786-791.

68. S Madhwani, J Vardia, P B Punjabi, and V K Sharma. Use of fuchsine basic: ethylenediaminetetraacetic acid system in photogalvanic cell for solar energy conversion. Proc. IMechE Vol. 221 Part A: J. Power and Energy.

69. Kiang, Daibin, Ito, Seigo, Wenger Bernard, Klein, Cedric; Moser, Jacques-E; Gräetzel Michael. High Molar Extinction Coefficient Heteroleptic Ruthenium Complexes for Thin Film Dye-sensitized Solar Cells. J Am Chem Soc. 2006, 128(12), 4146-4154.

70. Arturo Solis-Herrera1, Maria E. Lara2 and Luis E. Rendon3& PHOTOELECTROCHEMICAL PROPERTIES OF MELANIN. Nature Precedings : hdl:10101/npre.2007.1312.1 : Posted 12 Nov 2007.

71. Richard H. G. Baxter*, Nina Ponomarenko*, VukicaS˘ rajer†‡, Reinhard Pahl‡, Keith Moffat†‡§, and James R. Norris*§Time-resolved crystallographic studies of light-induced structural changes in the photosynthetic reaction center. 5982–5987 _ PNAS _ April 20, 2004 _ vol. 101 _ no. 16

72. Dadachova E, Bryan RA, Huang X, Moadel T, Schweitzer AD, et al (2007) Ionizing Radiation Changes the Electronic Properties of Melanin and Enhances the Growth of Melanized Fungi. PLoS ONE 2(5): e457. oi:10.1371/journal.pone.0000457

73. Ling, M.M. & Bao, Z.N. Chemistry of Materials 16, 4824-4840, (2004).

74. Grätzel, M. Photoelectrochemical cells Nature, vol 414, 338-344 (2001).

75. C. C. Felix., J.S. Hyde., T. Sarna, R.C. Sealy, - Interactions of melanin with metal ions. Electron spin resonance evidence for chelate complexes of metal ions with free radicals, J. Am. Chem. Soc. 100, 3922-3926, 1978.

76. P. Meredith, B. Powell, J. Riesz, R. Vogel, D. Blake, S. Subianto, G. Will and I. Kartini, in Artificial Photosynthesis: From Basic Biology to Industrial Application ed. A. F. Collings and C. Critchley, Wiley, London, ISBN: 3-527-31090-8, 2005, ch3; J.E de Albuquerque, C. Giacomantonio, A.G. White and P. Meredith, App. Phys. Lett., 2005, 87, 061920.

77. P. Meredith and J. Riesz, Photochem. Photobiol., 2004, 79(2), 211-216.

78. T. Sarna, B. Pilas, E. J. Land and T. G. Truscott, Biochimica et Biophysica Acta 1986, 883, 162-167.

79. M. M. Jastrzebska, H. Isotalo, J. Paloheimo, and H. Stub, J. Biomater. Sci. Polymer Edn. 1995, 7(7), 577-586.

80. M. A. Rosei, L. Mosca and F. Galluzzi, Syn. Met. 1996, 76, 331-335.

81. Meredith, P., Powell, B.J., Riesz, J., Nighswander-Rempel, S.P., Pederson, M.R., and Moore, E.G. (2006). Towards structure– property–function relationships for eumelanin. Soft Matter 2, 37– 44.

82. -Kaur, B., Khwaja, F.W., Severson, E.A., Matheny, S.L., Brat, D.J. and Van Meir, E.G. HIF1-_ in glioma angiogenesis. Neuro-Oncology. J. 4, 134-153 (2005)

83. Marmor, M.F., Wolfensberger, T.J. The retinal pigment epithelium. Oxford University Press 1998. Pp186-187.

84. Forest, S.E., and Simon, J.D. (1998). Wavelength-dependent photoacoustic calorimetry study of melanin. Photochem. Photobiol. 68, 296–298.

85. Paul Meredith and Tadeusz Sarna, The physical and chemical properties of eumelanin; Pigment Cell Res. 19; 572–594, 2006.

86. Brattain, W. H. & Garrett, C. G. B. Experiments on the interface between germanium and an electrolyte. Bell Syst. Tech. J. 34, 129–176 (1955).

87. Gerischer, H. Electrochemical behavior of semiconductors under illumination. J. Electrochem. Soc. 113, 1174–1182 (1966).

88. Kalyanasundaram; K. Photoelectrochemcial cell studies with semiconductor electrodes: a classified bibliography (1975-1983). Solar Cells 15, 93–156 (1985).

89. Licht, S. Multiple band gap semiconductor/electrolyte solar energy conversion. J. Phys. Chem. 105, 6281–6294 (2001).

90. Fujishima. A. & Honda, K. Electrochemical photolysis of water at a semiconductor electrode. Nature 238, 37–38 (1972).

91. Brabec, C. J. & Sariciftci, N. S. Polymeric photovoltaic devices. Mater. Today 3–8 (2000).

92. Wöhrle,. D. & Meissner D. Organic solar cells. Adv. Mat. 3, 129–138 (1991).

93. Shaheen, S. E. et al. 2.5% efficient organic plastic solar cells. Appl. Phys. Lett.

78, 841–843 (2001).

94. Tuladhar, D. et al. Abstract, Int. Workshop Nanostruct. Photovoltaics, Dresden, Germany <http://www.mpipks-dresden.mpg.de> (2001).

95. Grätzel, M. Perspectives for dye-sensitized nanocrystalline solar cells. Prog. Photovoltaic Res. Applic. 8, 171–185 (2000).

96. Savenije, T. J., Warman, J. M. & Goosens, A. Visible light sensitization of titanium dioxide using a phenylene vinylene polymer. Chem. Phys. Lett. 278, 148–153 (1998).

97. Thathachari, Y.T., and Blois, M.S. (1969). Physical studies on melanin II. X-ray diffraction. Biophys. J. 9, 77–89.

CPSIA information can be obtained at www.ICGtesting.com
Printed in the USA
BVOW05s2018050116

431887BV00011B/165/P